B. D Nightingale

Practice in Weaving and Loom-Fixing

A complete manual for the weave room

B. D Nightingale

Practice in Weaving and Loom-Fixing
A complete manual for the weave room

ISBN/EAN: 9783337105662

Printed in Europe, USA, Canada, Australia, Japan

Cover: Foto ©berggeist007 / pixelio.de

More available books at **www.hansebooks.com**

PRACTICE

IN

Weaving and Loom-Fixing.

A Complete Manual for the Weave Room.

With full detailed instructions respecting the
Construction and Operation
of Woolen and Worsted Looms, including
necessary calculations.

By B. D. NIGHTINGALE,
Weaving Master.

Published by
THE TEXTILE RECORD,
425 Walnut Street, Philadelphia.
1887.

Price, 75 Cents.

TABLE OF CONTENTS.

3

CHAPTER VI.
STARTING A WARP.

CHAPTER VII.
WARP MATTERS.

CHAPTER VIII.
SHUTTLES, TEMPLES, AND BELTS.

CHAPTER IX.
IN THE WEAVE-ROOM.

CHAPTER X.
CALCULATIONS.

CHAPTER I.

MAKING THE LOOM READY.

PREPARATORY OPERATIONS—THE BOXES—FITTING THE PINS
—THE SHAFT—THE PICKING PLATES—THE PICKING-BARS
—SETTING AND FASTENING THE SHOE—MAKING CONNEC-
TIONS—THE SWEEPSTICK AND STRAP—THE CRANK-SHAFT.

Fastening the boxes.—If I were to begin the overhauling of a loom, I should strip it of everything but the two principal shafts, and those I would lift out of the boxes. Fixers have always been troubled with the boxes getting loose, and the difficulty that we experience in reaching them to tighten them up makes adjustment of a loose box a job that any fixer dreads. So that now, while everything is out of the loom, it is a good time to make them permanently fast. On the new Cromptom looms the rib that is cast on the frame helps to keep them in their place; but the old looms are always a trouble if the boxes are not well fixed.

Take the box out, and lay a straight edge on the inside. If the middle rib is higher than the outside ones, two of the best bolts you can find will not keep it from rocking on the frame; and unless you have one or two unsightly pieces of picking-stick to brace it down, the chances are that you will have to crawl under the loom very often to tighten the box and bark your knuckles. File across the ribs until your straight-edge shows them to be even, care being taken lest you get them slanting from either side.

Fitting the pin.—In putting the box on, take time enough to fit a good pin through both box and frame, and in such a manner as to draw the box down, so that it fits snug on the top of the frame. This is often carelessly done, and the pin, when holding the box up too high, prevents it from obtaining a bearing on the top of the frame as well as the side. If this is the case, you cannot keep it tight, no matter how hard you tighten up the bolts. Do the job well, taking time enough for it, and you will save yourself much disagreeable duty.

Having fitted the box to the frame, so that it has an even bearing on both top and side of the frame, the bolts snugly tightened will hold the box so that it cannot rock or start.

Set-screws.—Before putting the boxes on the frame, look after the set-screws or cap-bolts. On a loom that has been run any length of time they are usually found worn out. It takes but a very short time to spoil the set-screws in the bottom shaft boxes, if they are allowed to rattle around loose until you have been compelled to tighten them up to keep the shaft in its bearings. While you are at it, fit in good set-screws. Let them screw in rather tight. You cannot screw them in with your fingers, perhaps; but when they *are* in, you can take a large wrench and put your muscle to it, and the caps will stay there.

The shaft.—We will next turn our attention to the shaft. The big gear should have no worn-out teeth in it. They always come where the pick begins, or the protector strikes. If they are worn at all, I should cut a new key-way one-quarter of the way around the hub. The gear should fit to perfection. We want the picking shaft to stay in its place when we get through with it, and we cannot afford to leave the gear-wheel running out of true on the shaft.

So, if not perfectly true, get it fixed, and, when it is keyed on, see to it that the key fits snugly the whole length of the key-way.

The advantages of a perfect-fitting key are that there is no danger of the hub being split in driving it in. A light riveting hammer will drive it in, and it will stay for good when it gets there. And, not the least important, you do not have to waste several hours of precious time to get it out when you want to.

The picking plates.—Next look after the picking plates. If they have ever been loose, the key-way will be found ragged, and almost useless. Either cut a wider key-way, or turn the shaft over and cut a new one on the opposite side. Be sure that each plate is fastened on the shaft exactly like the others and spare no pains to get this result, or else trouble will follow, and be likely to continue for a while, too.

While the plates are loose, see if the grooves for the picking-roll-extension are not worn too badly on the edges. If they are, it is very important that they be fixed, either by planing or filing, while you have a chance. If the grooves in the plate, and the ribs on the extension, are not square on the edges, the bolts are broken sometimes at the rate of two or three in a day. The usual treatment is to put the biggest bolt that it is possible to get in through the plate, and if that breaks, to do the same thing over. The right way is to fix the grooves and the extension, and then it does not take much of a bolt to keep them fast. Therefore it pays to make a good job while the plates are off.

Before putting on the plates put a good collar on the shaft, on the gear end, to prevent end-play. It is very important to do this, for it saves the gear more than anything you can do, by keeping it tight in its place and preventing it from slipping with the pinion-gear. Having done this, slide the picking-plates on, but do not drive the key tight until the picking bars are in place, so that you can try the roll on the shoe, and make it strike the latter in the right place.

The picking bars.—Before putting the picking-bars in place, have them straight. Little attention is paid to this matter sometimes, and a bar is put in while twisted and bent, and then the fixer wonders why he cannot get as good pick on one side of the loom as he does on the other, or why it is that one side takes a sweep-stick 19 inches long, and the other side one of 17 inches. A good pick, or at least a trustworthy one, cannot be obtained unless the picking-bar is straight.

Setting the shoe.—In setting the shoe we come to a task that often puzzles fixers of long experience, and the rules, as laid down by different authorities, are seldom alike. Some assert that $7\frac{1}{2}$ inches is the right place; while others cannot run a loom successfully short of 8 inches from the socket. Every man has an idea, and possibly a good one, too; but I do not believe in doing much measuring for the sake of following some rule that I have read of. It is far less important than studying out the *principle* and being guided by that. Any fixer will find this out, if he ever gets hold of a kind of loom where the picking-shaft is farther from the back-girth upon which the socket rests, than it is on the

Crompton or Knowles looms. It will then be found that the 7½ inch rule is not the thing. In that case we are obliged either to try until we get it right, or to learn the principle upon which the picking-shoe is constructed, and work from such knowledge.

The distance of 7½ inches is, in my opinion, too far back to set a shoe on a Crompton loom. The pick loses its force at the point where it is most needed. The shoe is so constructed that from the bottom the incline is about one-half as steep, for about half the distance travelled by the roll, as it is from the middle of that distance to the top; so that when the roll strikes the shoe at the bottom, the speed of the picking-stick is comparatively slow at first, getting swifter and stronger as the roll reaches the steeper parts, until the point is attained where the shuttle leaves the box. Now, if we set the shoe back, the relation to the shoe, of the circle described by the picking roll, is changed completely, and, if anything, it gives more power at the beginning of the stroke than it does at the last, and you borrow power by lowering the lug-strap.

You then have a pick that jerks the shuttle full speed at the start, making it easy to fly out on the slightest provocation, and sometimes without any provocation at all. Furthermore, the lower the lug-strap is set, the nearer you get to the wrong end of the lever, and it takes an immeasurable amount of increased power to throw the shuttle—a loss every way.

To move the shoe too far forward is to go to the other extreme from setting it too far back; the only difference being that you cannot run the loom at all if the shoe is set directly under the shaft. Therefore, to get the best results, I think 7¾ inches is the right place to set the shoe on the Crompton loom; and I state this measurement, not on account of the inches it takes, but to show that if you should set other things right, and then put your shoe where it would do the most good, you would find it about 7¾ from the socket.

Fastening the shoe.—Having become satisfied that this is the place where we want the shoe, then measure every time, and in putting our picking-bar in place make the shoe fast enough to stay where you put it. If it is a new shoe, file it a little inside. Most fixers put the shoe on, if it will go on, without filing, and a bur in the corner gets all the strain of the tightened set-screw. When working, the bur soon gets crushed and the shoe is loosened. If it be filed square and true inside, so that the ribs rest solid on the bar, you have a bearing which will hold better than an over-tightened set-screw, and you seldom burst the shoe in fastening.

Fastening the picking-arm.—On the same principle I would fasten the picking-arm. I have often found home-made picking-arms with no rib around the edges, on the inside. No part of this would touch the bar but the middle, which in fact should never touch at all, and you could turn the set-screw as tight as possible and then rock the picking-arm on the bar. It only runs but a short time before it is loose; and as the fixer dislikes to take it to the bench to make a good job of it, he gets under the loom, tightens it up, perhaps several times a day, until he gets sufficiently disgusted. Then he takes it to the bench and finds the bar worn so that no picking-arm, however perfect, will fit on the bar and have a true and solid bearing. It is run in this way for years, and, times without number, some one has to fix that picking-arm. On the new style looms, especially, careful attention should be paid to this matter. If the inside of the picking-dog is not filed carefully, to square it, the bearing on which the loose picking-arm works soon gets cracked and broken.

Making connections.—Having got everything on the bar, put it in place and proceed to make the connections. The picking-arm-stud should be in good shape if you do not want to be fixing it every little while. I would not put a stud on unless there is a good enough thread on it to permit of the nut being screwed up very tight. If the stud should get loose for a few minutes, when you go to fix it you will find it spoiled. It should never get loose; and it will not if it has a square, solid bearing on the picking-arm.

The pin in the stud.—A very insignificant but troublesome thing is the pin in the stud. It is too small an affair to be worthy of much care, so the fixer twists a piece of wire in the hole to serve for a pin and, of course, it cannot come out. He cannot get it out sometimes when he wants to put a new sweep-stick on in place of that which this same crooked piece of wire has spoiled, by gouging out the stud-hole of the stick.

I regard it as a matter of great importance to look after these so-called little things, and when I put a stud on, I cut a leather washer that will fit tight on the stud, and then drive a tight, straight pin in the stud. If it fits the hole all right, it will not come out; but if it is too tapering it cannot get a bearing excepting at its thickest part and of course will come out. Fit it in right. This will not take long, and it will save you many times the labor expended because the job will last, to say nothing of the sweep-sticks saved.

The sweep-stick.—The sweep-stick should always have a rivet or bolt in the end to keep it from splitting; and the prudent fixer or overseer will see that this is done and a supply of them already prepared before they are needed.

The sweep-strap.—For a sweep-strap, which is the next thing we will put on, various kinds of material are used. I do not have a very high regard for rotten belting for this purpose. Nor are the canvas sweep-straps without their faults. The canvas straps, if everything is in the best of order, are the most desirable; but frequently they break the screws in the stirrup-strap until the picking stick is so full of screws that you are compelled to take it off for want of room to put in another screw. Some fixers put the stirrup-strap on the side of the picking-stick. It is a slovenly and undesirable way of doing it. If the screws are breaking it is the fault of the loom. They will not break if other things are running all right. It generally occurs when the pick is such as to compel you to put the lug-strap too far down, and that is never necessary. If power enough cannot be obtained, overhaul the picking-motion and make things right; and if other things in connection with the picking-motion are doing their work properly you can raise your sweep connections on a level and there will be plenty of power. Do not punch your sweep-strap full of holes; one hole is enough.

In making our sweep connections, now that we have had everything off the loom, I should have the picking-stick-stud in the centre of the slot; and then fasten it temporarily until I could try the rolls on the shoe. You can now move the picking-plates so that the roll strikes the shoe all right, and then make them fast. This done, try the sweep. The picking-roll should touch the shoe at the bottom, and give enough sweep to bring the picking-stick to within one inch of the bunter. If it does this, make the connections fast and you will not have to move them much when you come to start the loom.

The crank-shaft.—We will now fasten everything in connection with the bottom shaft and we are ready for the crank-shaft. The boxes should be given the same thorough overhauling that we gave the bottom shaft-boxes. In gearing them together turn the picking-ball so that it just begins to move the shoe and then gear the two shafts together with the crank not quite on the top, or, inclined one tooth toward the lathe. We may have to move it, but we will try the lathe first and see if we are right.

The picking motion on the Knowles loom is practically the same as on the Crompton, and the rules for running the latter, apply to the Knowles. In point of construction it is not excelled by any loom made. The same, in fact, may be said of any picking-motion where a shoe is used; the only difference being in the length of the shoe-shaft and the distance from the back of the loom.

When a Knowles loom is sent from the shop you will find that the sweep-stick is in two parts, and bolted together midway. It is a convenient arrangement and can be used on any loom to advantage.

There are other slight changes on the Knowles picking motion, and, taken as a whole, it is a most excellent arrangement.

CHAPTER II.

LATHE AND SHUTTLE-BOXES.

CRANK-ARM OF THE LATHE—SETTING THE LATHE—FIXING THE SHUTTLE-BOXES AND FITTING THE SHUTTLES—BENDING THE BOX-ROD—PICKERS AND PICKER-RACES.

Crank-arms of the lathe.—In beginning work on the lathe we will first put on the crank-arms. The strap that goes on the crank, whether it be of iron or leather, should not be tight. It is very easy to make a loom run hard if this is too tight, more especially with an iron strap. The thickness of the crank-arm should equal the diameter of the crank. If it does not, when the bolts are tightened up the band or strap squeezes the crank so hard as to make it almost impossible to move the loom by hand. This applies to new looms mostly.

Setting the lathe.—Having put the crank-arms on, we will next set the lathe. The race-board should be five-eighths of an inch below the breast-beam, and leveled before fastening. The middle sword should be the last to be fastened, and should pull down on the lathe. To accomplish this, tighten up the bolts on the bottom just enough to hold what it gets. Then place a block of wood on the race and strike with a loom-weight. One or two blows will sag the lathe down enough to hold, when the bolt should be tightened up as hard as possible, so that the jar of the loom will not let the sword slip up.

Shuttle-boxes.—We can now find plenty to do on the shuttle-boxes, whether they are new or old. If new ones they should be taken out and filed, as the corners and edges are usually found to be rough and sharp. Do this job thoroughly, and do not be in too much of a hurry, as you will never have a better chance to put the boxes in proper shape. I find it a very advantageous thing to file the sides of the mouth of each box, the lower half the most, as it will let the shuttle touch the top of the side, before it will the bottom, and prevent filling-cutting.

Have a shuttle handyl to try in the box while you are filing, and you can get each box so that the side of the shuttle below the eye cannot touch the lower part of the box-side at all, and you will find that it prevents filling-cutting at that spot very effectually. File the edges of the long slot in the back of the box through which the picker slides. File the sharp edge off only, and do not neglect it, for the shuttle is liable to be damaged if this is not done, and then it causes the shuttle to bind tighter than is natural, and the fixer wonders why he cannot get spring enough off of the swell to let the shuttle go in easy. For the same reason the boxes should be polished with emery cloth. The gum and rust on the inside of the boxes bind the shuttles too tight, and everything about the picking-motion has to be run a little different while they are in this condition. After they are worn smooth the fixer has to go over them and tighten up the binder-springs. Perhaps he has to bend the swells a little differently, and fuss with a new loom for many days before it is "limbered up," as they term it. I prefer "limbering" the loom up at the start, so that I have less altering to do.

Many fixers have great trouble in making the boxes work on a new loom, when the principal trouble lies in the fact that the rusty or gummy box causes the shuttles to stick, and they hold into the picker, or on the iron slide or evener. This is a small matter, and some may be disposed to pass over it lightly, but if fixers will be more careful in starting a new shuttle-box, and will clean it and polish it up before putting it in, it will be found a very profitable thing every time.

Fitting the shuttles.—Having cleaned and polished the box, before putting it in the loom I would fit the shuttles. In this part of the work no fixer needs to be reminded of the importance of care, but I think very few realize how much trouble they might save in the running of a loom if they put all the skill that care and study would enable them to do, when bending and fitting the swells. We know that it is a common occurrence for the shuttle in one box to work differently from the rest. If we have a shuttle that lags a little we find that others from the same side have enough power. A crooked shuttle is nearly always thrown from a certain box; the same with a shuttle that flies out after passing through the shed. Of course, this is not always the case, but most fixers will agree with me that in the majority of instances certain boxes give more trouble than the rest.

The reason must be that they are not alike; and of course, the difference must be in the bending of the swells or binders. In bending them, therefore, I would use the utmost care to have them exactly alike in every way.

The binder should touch the shuttle for a space of about 5 inches, 2½ inches on each side of the place where the binder-spring touches the binder. If one swell touches most, nearer the back end of the spring, and another in the same box touches most, nearer the front end, there is a vast difference in the pressure that is brought to bear on the shuttles in those boxes.

My reason for bending the swell so that the contact with the shuttles comes on each side of the point where the spring touches, is that to put it nearer the back end of the swell, binds the shuttle too tight and does not allow the slides, that even up the points of the shuttle, to push them forward easily. If this is the case, the box is held up by the point of the shuttle resting on this slide, and will cause the picker to catch in the box. On the other hand if the swell is bent so that the contact comes on the front end, it does not hold it tight enough.

In bending them, round the neck of the swell so that the shuttle does not strike a corner as it enters the box. Some believe that is the best way, but most fixers do not. The side of the shuttle is soon worn away, unless the swells are bent so that the shuttle has an easy entrance in the box.

While you have your box at the vise put all your swells in and try one shuttle in all the boxes and look through each box to see if you have not twisted the swell in bending it. If it is twisted you can see that it touches on but one side, and consequently you do not have more than one-half the surface in contact with the shuttle that you would if both sides touched. One would hardly suspect this, and for this reason I like to bend the swells where I can get at them and see to do it right. I like to take pains with them and when I am through with them I do not bend them every time I go near the loom, or allow any one else to do it, either. I cannot bend a swell properly with a loom weight or an old picking-ball, and I do not think any one else can. There are many "special" bends for various troubles, and I have tried nearly every way, but they are all unnecessary. Bend the swell right, and if the shuttle flies out or is thrown crooked, fix the right thing. You will have less trouble in the end.

Bending the box-rod.—In putting the box in the lathe there are a few points about bending the box-rod, not generally noticed. The back of the box should tip up about one-sixteenth of an inch, and to do this the box-rod has to be bent. In performing this operation, the bend is made at the lower end of the rod, or where it passes through the hole in the picker-stick-stand. When the box is lifted to the fourth one, a difference of $\frac{1}{4}$ of an inch may be found. The same thing occurs sometimes when the rod is bent towards or from you, as you face the box. In this case it will bind the box when it is lifted up, and I have known fixers to file the shuttle-box-guides, and even the box, when the fault really lay in the way the rod was bent. The remedy, therefore, is in bending the rod in such a manner as to leave it perfectly straight where it passes through the stand below, and you can depend upon having each box come up, tipped 1-16th of an inch, if you set them that way; if 1/8th, they will all be alike.

Pickers and picker races.—We will now put the boxes in the loom without putting the box lever connections on, and turn our attention to the pickers and picker-races.

CHAPTER III.

THE SHUTTLE.

On a new loom there is not much to do but to put the pickers on, and see to it that we have a packing solid enough to keep the picker from going too far back, so that the shuttle will not catch. On nearly all looms there is a chance to improve a little in this, so we will consider that we are working on an old loom.

We often find that a loom readily makes trouble by throwing the shuttle out, and also by throwing a crooked shuttle. It perhaps cuts the filling, and when we have a very heavy warp in, it will bump the shuttle against the top of the box and rough it up by so doing. On some warps we have no trouble, while on others we have to fix the loom every day. We then blame the reed or the heavy warp for the trouble, but at the same time we know that on other looms the same reed and warp will cause no trouble. To get such a loom as that in such a shape that we can put anything into it and not be afraid of it, is a very desirable thing to accomplish, and I think it can be done. I have had looms that I dreaded to be called to, and I always had trouble if the warp was in any way difficult. By studying out its peculiarities and having a "brush" with it when I had the time, I have got such looms into the traces again, and every warp would run right along. I usually found the trouble in the way the shuttle-box, in connection with the picker-race, was lined. Sometimes the whole lathe-end would be found out of line, and it looks like such an important job that few fixers would loosen up the lathe end to fix it.

To find out how we stand let us put a good reed in the lathe and put a straight edge on it, extending into the shuttle-box. The straight edge should be perfectly true and in using one for this purpose, do not take anything for granted, but find whether you are using a really straight edge or not.

In trying it in the box and against the reed, if everything is all right, the straight-edge will be one-sixteenth of an inch from the face of the reed at the edge or side of it. If it is any more than that or any less, it is not as it should be.

Suppose you find that the straight edge

touches the reed and is, say, one-sixteenth or one-eighth of an inch from touching the box at the back end of it. One way of remedying this is to loosen the shuttle-box-guide and pack it so as to bring the box out far enough to touch the straight edge and throw it one-sixteenth of an inch from the reed. This will do very well so far as the box is concerned, and it is the way this job is usually done, but if you bring the box forward and not the picker race, how is that going to keep the race in line with the box? And any loom-fixer should know, if he does not, that the picker-race should run in exact line with the box. If not, the point of the shuttle touches the picker in one place when it is back in the box and if brought forward and the shuttle is held against the reed as it should start out, it will be found that it touches the picker in a different place; and when thrown out quick, the shuttle does not have time to swing back against the reed and consequently goes through the shed in a sideling sort of a way. Perhaps it is thrown outward on the race, and in bounding back on the reed is thrown out, after passing through the shed. These are our crooked shuttles and our shuttles that fly out after passing through the shed; the worst of all things to fix. In order to avoid them we must have the picker-race parallel with the box; and if it is an iron-end lathe, the packing out of the shuttle-box guides will not help matters much, but will be more likely to make them worse.

Causes of difficulties.—If the back of the shuttle-box is found to be farther back than it should be, before doing anything ascertain first whether the race and box are parallel. Sometimes the front of the box is too far out, or perhaps the front shuttle-box-guide is packed too far out; or if not, the casting is too thick. The front shuttle-box-guide is not apt to be too thick. It should face out even with the inside of the box and usually it does; but often a piece of pasteboard or leather has been slipped in behind it. But, if everything is all right in regard to the front end of the box, the trouble with the back end is in the whole lathe-end being out of true.

Now tip your box up, so that the back of it is one-sixteenth of an inch higher than the front and then see if the slot in the picker-race is right as regards being level. It should be level and thus allow the box to be higher at the back than the slot is. If not found correct you can alter this part, while you have the lathe-end loose for bringing it in line with the reed.

Having become satisfied which way you want to throw the end, loosen up all the bolts and lag-screws and pack with paste-board. It does not need a great quantity, as you will find out, but put enough in so that you can tighten up the bolts very tight and yet not throw the end out of line. If, in looking at your boxes and race, you found that the front end of the box and the face of the front guide were out farther than the reed, which often occurs, it might be enough to put one thickness of press-paper nearest the sword and two nearest the middle of the lathe. That will serve to throw the sword backward, and the end of lathe forward, which will perhaps bring it far enough to make the back end of the box 1--16 of an inch farther out than the reed, which is what you are trying to get. If the end needs tipping either up or down, do it now while you have it loose.

Of course to make a success of the job a great deal depends upon the judgment of the fixer, and he should proceed with confidence, remembering that there is a right way, and it only needs perseverance to get it.

After fastening the lathe-end securely, try the box in again and draw the picker out slowly, allowing it to push the shuttle before it. Hold the shuttle snug against the back of the box and against the reed, and see if the point of the shuttle keeps in one place on the picker. If the lining of the lathe-end has been accurately done, and the box and race are parallel, the point of the shuttle will touch in the same place whether the picker be back, or forward, to the end of the stroke; and if you have got this result the loom will throw the shuttle straight; it cannot help it.

Other causes of trouble.—Of course there are other things that can make the shuttle work badly; but usually they are very simple things, and many fixers do some big overhauling to make a loom throw the shuttle straight, when some such thing as a bent reed is causing all the trouble.

Bending the reed.—Right here I would call attention to the practice of bending the reed back with the hammer when it faces out too far for the back of the shuttle box. I have just shown how the reed, box guide and shuttle box can be brought in line with each other, and I think it is a great mistake to spoil all the reeds in the weave-room on a few looms that are simply in need of proper adjustment. The trouble caused by this pernicious practice is iucalculable. I do not think these things would be ' so frequently neglected, if loom-fixers fully appreciated the importance of them.

Putting in the pickers.—While we are working at the picker-race we will put the pickers in. They do not need to be gouged out to fit the point of the shuttle. That practice is sometimes convenient to resort to if the loom is not in good order and we wish to make the shuttle follow the old rut ; but at best it is only a makeshift. If the box and race are parallel, the point of the shuttle will do all the gouging necessary, and in the right place, too.

Packing behind the picker.—The packing behind the picker, like lots of other things about the loom, can make plenty of trouble if not properly made. Bent box-rods, broken shelves · in the boxes, broken box levers, broken picking sticks, bent picking-bars, besides the smashes resulting from this general upsetting—are among the troubles caused by defective packing behind the picker.

I used to have an idea that to make a soft packing would help filling-knocking-off; but I find that it only takes a few picks to drive the packing back so that it amounts to nothing as a cushion. I now make a ball out of cloth and yarn, solid enough to last, and between this ball and the picker I put a piece of good leather, putting it so that it has a projection on one side to fit in the picker-slot, and thus be kept in place.

If any one imagines that to make a packing too solid causes the shuttle to bound, I would assure him that it would be working at the wrong thing to attempt to prevent shuttle-bounding by using a soft packing. It has to be hard enough to last; and if the shuttle bounds, fix that with the swell.

Putting in picking-sticks.—The picking-sticks can now be put on. Much could be said of the quality of the picking-sticks we usually get. The best are made of sound second-growth hickory. That we cannot always get them we know very well; but an overseer cannot do a much better thing for the general good of the weave-room than to see to it that a first class article is provided.

Before putting them on, a rivet or bolt should always be put in near the hole for the stud, and one just below where the stick strikes the bunter. The hole for the stud should be bored squarely through the stick. Some one might say that any one would know that; but *I* know that there are many who do not use proper care in boring the holes in the stick, and consequently the stick is twisted when put in. If, then, it binds between the race and the spindle, a jack knife is brought into play, and a not very scientific slice is gouged out of it. Now a stick cannot last if it is not properly treated, any more than anything else on a machine; and if fixers would use the care in working with such things that a good carpenter would, they would have more time to think over their work, and keep the looms running at the same time.

The bits employed are usually so dull that they are not fit for boring wood as hard as hickory is. In crowding a dull bit into the picking-stick at the top there is danger of splitting the stick; it may be very slightly, but when it is put on the loom the screw gives trouble by coming out, and in trying to put a larger one in the stick is ruined, Thus a new stick is spoiled where it may have lasted for months. The cost is not much, of course; but the demoralizing effect of a fixer having to keep on a run to fix these simple things amounts to something in the aggregate, and should be prevented. Use good bits, and take care of them as a carpenter would. They are just as precious to a loom-fixer as they are to any one else.

The bunter.—Having put the picking-stick on, with the previously made sweep connections fastened to it, try your bunter. The bunter for the stick should be thick enough to hold it off the picker. Be sure of this, as this business of putting on new pickers and sticks amounts to a great deal in a fixer's work, and if the picking-stick is stopped.by the picker, neither of them can last long.

There are few things about a loom that are so sadly neglected as the bunter. It is too insignificant, in the minds of most loom-fixers, but this is a mistaken idea. The bunter should be made out of something that will last. I have tried rubber, and like it, but it does not last long enough. I think good leather put in tight, with the edges cut true, so that you can drive it in solid, is the best thing you can use.

The packing on the spindle.—The packing on the spindle is mostly a piece of picker. It does very well, but I think it is too hard to use for that purpose. A few pieces of good, hard leather answer the purpose better, as they do not batter the picker.

The protector.—Before leaving the lathe, let us look at the protector. It is often found bent at the ends. If it is bent, take it out and straighten it. The rod cannot have that easy swing that it should have, if it is not straight. Often the ends where the set-screw of the protector-finger fastens, are so cut up with the screw that we cannot set the finger where we want it. If very bad, they should be either upset, so that the holes can be obliterated, or a new piece welded on.

Have the daggers dressed, being careful not to get them too sharp. To get them the right length it is necessary to try them in the loom. The one on the dead side should be long enough to hold the other from binding the knock-off-lever tight. They should also be made to fit into the grooves of the knock-off-lever and the bunter. Often they do not, and the result is that looms will make smashes which, when tried, seem to protect perfectly. If the dagger on the lever side fits into the groove of the lever, and the one on the opposite side is a little high, when the loom is running the dagger on the dead side will strike the upper side of the groove in the bunter and glance with such force as to throw both of them clear of the grooves and thus fail to protect. If they strike perfectly even, nothing will be likely to bend them, and you can depend on them.

When you put the protector in, put the fingers on, and before fastening very tight, adjust the daggers to fit the grooves. Then make the fingers as fast as the set-screw will allow. In tightening the spring on the rod, be careful not to put too much spring on. This is a very common error. If the rod is straight, and works free, it does not require much spring to keep the fingers back on the swell when it is needed. The rod should have an easy swing, which will allow the shuttle, in entering the box, to throw the protector so as to clear the grooves in the lever and bunter.

Objections to a tight spring on the rod.—
I have seen looms running under fixers who believed in running the loom with a tight spring on the rod, and have closely watched the effect for many months, and I found it anything but desirable. In order to prevent the loom knock-

ing off, the shuttle had to come in the boxes with terrible force, and even that did not keep the protectors from catching occasionally.

The evils resulting from such a condition would take up a good many pages of this book, if fully described. The one fact of having to increase the power, if followed up, shows that the binders have to be tighter than is natural. The shuttle must have greater force to start it out of the boxes, and that takes another large share of this borrowed power. The loom requires a much tighter belt, making it a very serious thing for the weaver—more important on a loom than on anything else that can be named--on account of the frequency of the starting and stopping of it. Then there is the increased strain on the picking-motion, causing unnecessary wear on those parts; and, worst of all, when the loom does knock off, the immense power exerted by the over-tight belt causes broken gears, broken crank-shafts and loose boxes, besides a general racking of the loom in all its parts. Try running the spring loose, and you will find it gives better results. I have dwelt on this, because I know of so many cases where the springs are run too tight, with the results named.

The knock-off lever.—While working on the protector see that the knock-off-lever is not bound when the loom protects. It should go far enough to knock the handle off, and then the dead side should hold the loom, so that the long shipper rod may be free to stop the loom.

The shipper.—The shipper should always work easy. On a loom that starts and stops easily a weaver can accomplish twenty per cent. more than he can otherwise; not simply because it takes longer to start the loom, but in acquiring an easy, comfortable motion, and being able to use the hand for something else when changing the shuttle; and where girls run the looms the fatigue caused by working on a heavy shipping loom discourages them not a little.

The shipper-fork.—The shipper-fork, if properly adjusted, helps in this matter more than any thing else. If any of the parts bind, the fixer can easily detect and fix them; but fixers do not always think of the belt. If the shipper-fork is set close to the pulley, the belt does not have time to follow the motion of the fork. Consequently the fork pulls on the belt, as it would if the shafting were not running. It pulls in a small degree, but enough to cause a resistance, which makes it hard to ship. The fork should be set four or five inches from the point where the belt runs on the pulley, and should be nearly at right angles with the edge of the belt to prevent turning.

CHAPTER IV.

HEAD-MOTIONS.

The pump-motion.—There are many kinds of head-motions, and in speaking on this subject I shall aim to take up those that are in general use. The oldest popular one is the pump-motion. It is a hard motion when a great number of harness is used, but for 12 or 16 harness, or less, it runs very nicely if it be properly adjusted. The great trouble with this head-motion is that it has so many joints and connections, and these, when worn, cause a great deal of lost motion and give that jerking motion to the harness which is so hard on the warp and on the numerous connections of the harness motion as well. To get good results from the pump-motion, the studs and connections, if badly worn, as some of them are, should be replaced by new ones. It is poor economy, in my opinion, considering the small cost of repairing, to run this head-motion with the parts so badly worn that the crank-stud gets ¼ of the way around before it begins to open the shed; therefore, if possible, if you have a head-motion worn out, fix it up in good shape.

Fixing the head-motion.—I remember very well the mystery that loom-fixers used to attach to the work on the head-motion, and I think it was two or three years before I dared to take one apart. I know now that the best way to get an understanding of it is to pitch into it and have it out at first. If you have every part of it loose, and you are ready to begin to set it, I know of no better way than first to set the stud or bolt in the crank-plate, about where you think you will want it; then set the evener on the outside rod, about ¾ of an inch from the top of it, the reason for this being that you may want a little room to move it. Screw it fast, being careful to keep it at right angles with the loom-frame. Then put in the brace-rod on the opposite side and divide the space where the thread is cut on it, so that you have room to move that, if necessary. Then put in a T.-jack with side jack on it. Now you can connect the knife-casting on the brace-rod; but before fastening the nuts slide it up on the pump-rod and screw fast. This will allow the casting to swing into a natural position. Then fasten up the nuts on the brace-rod. Having done this, draw the pump-rod down so that the T-jack is perfectly level. Then fasten the pump-rod. In fastening this pump-rod, bear in mind that its work is to pull down, so that the casting connected with the arm should be lifted or driven as high as it will go to take up the play that may be in the studs and connections. When fastened after doing this it holds the evener where you want it.

Next, set the lower evener on its rod as you did the first one, and set the lower knife-casting with knife already attached, while the evener is held up to the jacks.

When you have fastened this rod, the eveners should be [closed] on the jacks, not tight enough to bind them, but tight enough to take up the play that may be between the jacks and eveners.

Setting the knives.—Set the knives so that the outside or back-jack-hook comes within 3-16th of an inch of the knife and the front-jack within $\frac{3}{8}$ of an inch. Many fixers allow much more room than that, but it is unnecessary and is an injury to the easy working of the head The first movement of the head-motion is the. slowest and the sooner the knives take hold of the jacks the easier will be the movement of them.

Setting the stands.—In setting the stands that hold the T-jacks be governed entirely by the side-jack. It should be in a perfectly perpendicular position, and the top of it evenly divided between the two eveners. Be sure of this. The object of setting the T-jacks so that they stand level when the shed is closed, may now be seen. The ends of the jack, describe part of a circle and, of course, are holding the side-jack nearer to the chain when they are level than when open in either direction; so that, if they do not stand level, the jack is withdrawn from the chain too far on one stroke and pressed too hard on it on the opposite stroke, making the head very liable to make mispicks by slipping the jacks off the knives, or in binding the jack on the evener. For the same reason the side-jack should be perpendicular. If the T-jacks are set too far forward, they cause the side-jacks to stand in a slanting position, with the top nearest the chain. When the jack goes down to raise the harness it crowds against the chain, and in going up the jack is withdrawn from the chain. The philosophy of this can easily be seen, but it is not always heeded. Nothing will so cause the breakage of jacks as neglect of this; and if the jack is crowded against the ball, the chain may be slipped on the cylinder, or even turned one pick too many.

Look into this when you are troubled with mispicks and you will readily discover the cause of the mischief.

Setting the chain cylinder crank.—In setting the chain cylinder crank you find a chance for experiment. The object is to turn the chain as late as possible to avoid wearing the knife and the edges of the jacks, and yet to get the cylinder turned in time for the next pick. This can easily be done, though not so easily described; but do not set the crank so that the cylinder begins to turn when the lathe is back. The effect is to cause the jacks to begin to change before they are clear of the knife and sometimes they will be worn out in a single week. The loom is also very inconvenient for the weaver, as, in stopping the loom, should the lathe come forward in the least, and is then pushed back, a mispick is the result. They are bad enough in this respect at best. Do not let the hook fol'ow the chain cylinder too far. It is apt to turn two notches in the check-wheel. If it goes far enough to start the roll down into the check-wheel it is far enough. Many looms are run with two, and sometimes three, springs on the check lever. The loom was made with one and one is enough, These unsightly and unscientific encumbrances should be avoided. It makes it hard, and very disagreeable for the weaver, who has to turn the chain so frequently, to have to exert so much muscle in finding the pick. If the hook does not pull too far, and if the jacks work free'y so as not to bind the chain, it will not turn.

Harness wire and strap connections.—

There are many ways of making the harness wire and strap connections on pump motion looms. Some use wire from the side-jack to the long bottom-jack. I prefer a harness strap, and I think every one would if it were given a trial. The wire makes a hard and unyielding connection and I consider it especially hard on the harness. The best connection is to use a short piece of wire to connect the bottom end of the strap to the stirrup on the bottom-jack. One end of the wire should be made with a long hook to go through the stirrup, and the other end should have a short, nicely made hook to go in the strap.

Then, for the top end of the strap, have a hook to go through the leather ; and then when you have the loom in position so you can tell where to set the bottom (or underneath) jack, you can bend the end of the wire so that it will go through the hole in the side-jack. To vary the length of the connection, you can use the holes in the strap. What better arrangement could you have for changing jacks? You can unhook your straps and lay them in a common pile as they will fit on any loom. Round the corners off on the top-end of the straps, so that they will not catch one another in changing the pick.

Underneath, the wires on the harness should

be the same as on the bow-jack loom and the harness-hooks the same distance apart, so that the harness can go anywhere. The strap should be a flat one, with a neat wire-hook—not a buckle—to fasten with. You can draw each harness connection to the same tension easier than by any other way. The top straps are the same as the bottom ones.

The finger-jack loom.—The finger-jack, up-right-lever loom was, I believe, the next thing to the pump-motion head. There are plenty of them in use to-day, but they have long since ceased to be made. These looms give trouble by making mispicks; but the head-motion is very easy and if kept in good order will run very well. The principal causes of mispicks are very simple, too simple to be thought of some-times. Of course, the many points about the head-motion make it too easily gummed up and clogged with dirt; or, the parts rubbing against each other may stick and not permit the fingers to fall in their places. My experience with them teaches that if these looms give trouble look for little things as the cause.

Setting up the head.—In setting up the head little difficulty is experienced, as the parts each have a place and cannot very well be mis-placed. The same may be said of the slots for the studs and rolls; they cannot very well be set wrong. The rolls should bring the knives together so as to close in on the jacks evenly, both back and front alike. When the arm, which connects the rocker arm with the crank plate, is in place, and the stud on the plate is on top, the knives should be closed snugly against the jacks, but not tight enough to bind them. Do not be too sure you are right, but try them both back and front, to see if either side binds. The jack hooks when at rest, and the shed closed, should not touch the underside of the knife. That part of the hook that slips through the fingers can become bent and allow the hooks to get in various positions.

Hooks and fingers.—In running this head it is important to avoid letting the hooks or fingers touch anything that moves, at a time when they should be at rest. The sides of the head of the fingers get worn perfectly smooth and flat. When hugged together, this smooth surface, which would lead anyone to suppose to be just the thing to slip freely, often has the opposite effect. The oil becoming gummy, causes the two heads to stick together, a kind of suction. If the fingers are wiped and fresh oil put on, it will stop it; but to make sure that they do not stick and make an occasional mispick, I take them out and grind the sides of the heads slightly rounding, This prevents them very effectually from sticking.

The part of the hook that slides through the fingers should be carefully bent if it does not bring the hook in the right place on the knife. All of the hooks should stand alike when tried with the chain off. The same with the part of the finger that touches the rolls or balls. If the underside of the finger is worn flat and is nearly as wide as the balls on the chain, file the sides so that they have no corners on, and when the balls on the chain move sideways on account of being worn, there is less liability of them catching the wrong finger.

Sticking of the slide.—Sometimes when the shed opens pretty wide the slide of the hook sticks or cramps in the hole in the finger. This can be prevented by trying each jack. Take the front knife off and with the lathe back, pull each knife up as far as it will go and then feel the finger to see if it sticks. It is a good idea to file the inside of the finger-head always before they are put on as there is no yielding when one of them catches; something has got to bend or break.

The shed.—I always avoid making the shed larger on these looms, by giving more sweep to the head-motion. I keep sweep enough on it for 20 or 24 harnesses and run the straps as low as possible. The chain should be set so as not to turn too soon. If it does, it catches the hooks on the edges of the knives as on the pump-motion, and is also liable to turn or slip the chain on the cylinder.

The bow-jacks.—This head with slight alterations is fitted with bow-jacks and makes a most excellent motion. It is a very safe motion on difficult work, being less liable to make mispicks than the finger-jacks. If it does make mispicks, they can usually be attributed to the action of the chain, which, if not made to work smooth and easy, will give the fingers a jar just as the knives begin to open, and then one of the wrong fingers is liable to hook on the knives.

A cause of mispicks.—There is one cause for mispicks on these looms that is very obscure and I doubt if it is generally known. I have seen fixers work for days on a bow-jack loom which would make a mispick two or three times a day. Nothing can be more annoying than this, as if the loom made a mispick every five minutes one might detect it.

An example.—I was troubled with a loom doing this and I tried for two weeks to find the cause. I could, perhaps, have moved things until I hit it, but I was desirous of learning why it was that one of these heads, which seemed in perfect order, would make a mispick occasionally; so I altered nothing until I found some clew. The fingers would shake when the head closed, but the jacks seemed all free and I thought I had tried them all.

After being bothered with it until I felt like giving it up, I discovered that one of the jacks was a trifle wider than the rest, and the knives, in coming together, would strike it, and the jar gave them all an almost imperceptible bound. I filed the jack and made them all even while I was at it, and I never experienced trouble again.

Try each jack.—I always tried each jack when I put them in and made them of equal width and made sure that none of them were squeezed by the knives, and the result was that I rarely ever had one of these looms make a mispick unless something else was out of order which could be easily detected. The same thing applies to the 1880 and 1883 Crompton loom, and I know of large weave rooms using the 1880 loom in which this idea was tried and found very satisfactory. It is a very natural thing for the fingers to jar if the knives bump together on the jacks, but I know that many do not think of it. As with many other simple things, it is harder to attract the attention of fixers to little things like these than it is to tell them of some big job, and yet these are equally important.

The horizontal motion.—The horizontal motion is one that does not meet with favor among fixers. Where it is used on more than 12 or 16 harness it is hard on account of the manner of evening up the jacks. The eveners have to draw all of the jacks up, which makes the top and bottom connections pull against each other. This unnatural movement causes a great strain on the straps, as well as on the head-motion, and on heavy work a large number of breakages occur. To make matters worse, there are five straps for each harness; and being strained at every pick they soon break, and will keep one man busy on an ordinary section fixing harness straps. The head is easy to regulate, and there is plenty of room to make the shed as large or as small as you wish ; but the most profitable thing for the fixer to do is to turn his attention to reducing the strain on the harness connections to a minimum. If the straps give trouble by jumping out of the sheaves or pulleys, the only remedy is to fit blocks of wood over them. This is not a scientific way of doing things, but you have the satisfaction of knowing that principle which causes your straps to be strained at one part of the operation and to be very loose at another is impracticable ; so you are justified in using means which give the best results.

These looms will give very little trouble if the work is not heavy and the number of harness small.

Another head-motion.—The 1880 head on

the Crompton loom is one that gives general satisfaction to all. Of course some persons can find things about it that don't suit them, but there are many who are thankful that most of the perplexities incident to the old style head-motion are removed, especially when so much more care is required in the production of woolen goods. The action of the head is practically the same as in the old style bow-jack looms; but it is obtained in a different way, which is so simple as to need no explanation to those interested.

Binding of the jacks.—The remarks regarding the binding of the jacks on this loom may be called to mind. The time of turning the chain-cylinder should be carefully attended to. The edges of the jack-hooks will be worn off in a few hours if they are suffered to bear on the edges of the knife in changing. Should they become worn through oversight, it is better to take them out of the loom at once and file them into proper shape. They should be filed with a three-cornered file to get the proper bevel on the hook. Be careful to make them all even. Grind or file the knife to proper shape also.

The chain-cylinder.—The chain-cylinder should turn towards the loom. I have seen fixers turn it from the loom rather than take the chain off and turn it to make the twill run right. When the chain is in place the bar does not stand on the cylinder as on the old bow-jack loom, but is a little past the centre on the side farthest from the loom. The jack-fingers are curved slightly to suit this condition. Now, if the cylinder turns from the loom, the ball on the chain lifts the jack-finger slowly, and is too long in getting it into its place, so that we have to set the time of turning or starting too soon for the knives to get out of the way of the hooks on the fingers.

On the other hand, if we turn the chain towards the loom, the ball does not come under the finger until nearly in its place, and when it does touch the finger, it lifts it quickly. Consequently we can so time the turning of the chain that the knives have ample time to get clear of the hooks before the finger begins to rise. There is another reason for turning towards the loom, of which mention will be made in connection with the filling-chain farther on.

Special head-motions.—The Knowles head-motion and the Crompton box-motion are the same movement in principle. The cylinder gears turn the vibrator gears when the latter are brought in contact with them, and, like the Crompton box-motion, a vibrator gear, once turned, remains in that position until it is geared into the opposite cylinder gear, hence the open shed. It may not be proper to say that the Knowles is like that of the Crompton box- motion, but instead, to reverse the expression, because the credit of bringing that excellent idea out belongs to the Knowles. The whole mechanism of the Knowles head-motion, for beauty of action, and thorough construction is not excelled.

Of course there are different opinions in regard to the open-shed principle, but the popularity of the Knowles loom is strong evidence that the time is near at hand when those who favor the open-shed principle will be quite as numerous as those who oppose it. I regard it as a very fortunate thing that we have the two different ways from which to choose. I will not intrude my opinion as to which is the better way, but will leave it for those who use the looms to decide, from the teachings of experience.

Method of operation.—In running this head-motion the fixer will have very little trouble. It is so constructed that most of the parts cannot be misplaced. The part of the cylinder gear that operates the harness is adjustable, and therefore can be moved forward a few teeth from the part that operates the box-motion. This allows the shed to close on the pick as it is beaten in by the lathe. Some set it 6 or 7 teeth forward of the box-motion; but this is too risky. To move it too far causes the shed to close on the shuttle, and thus to fray the warp on the sides. Great care should be used in setting the shells that they

be both set alike, so that they engage the vibrator-gears at the same time. The lock-knife
should close in on the vibrator-lever at the same
instant that the gears begin to move. If not,
they are liable to be slipped.

Strapping the harness.—Avoid strapping
the harness too tight. There is no need of it;
and unnecessary friction is added thereby.
Should the soft set-screw ever need replacing in
the chain-shaft-gear, be careful to get it in the
same position as before, for should the chain turn
too soon, it will let the vibrators drop too soon,
and cause the gears to slip out of the cylinder
gear.

Sweep of the boxes.—The sweep of the
boxes may be adjusted by the bolt and slot in
the compound levers. If the box chains are
worn in some places more than in others you
cannot bring them to the right height, of course.
If this is the case, do not botch up the chains,
but take them off and file them even. If the
work does not call for the fourth box, I would
run the shuttles in it, even on plain work, enough
to keep the chain worn even.

Balls on the warp chain.—The balls on the
warp chain should be fixed to prevent the possibility of them touching two levers at once, a
thing that they can easily do. The fingers or
levers are malleable, and if new and old ones are

used together they should be made to come even by bending. A little care on the part of the fixer is all that is required to obtain satisfaction from this excellent head.

Harness on plain looms.—The operation of the harness on plain looms was formerly by means of cams and rollers. These have been superseded by the side levers which enables each harness to be operated separately, thus giving greater capacity for changes of weaves.

Cam-looms.—Minutely to describe the different "cam looms" would be superfluous, as, in principle, one means all. There is little to it on any of them, but, simple as they are, they need to be properly adjusted to get good results. The rollers should be adjusted so that the part over which the back-strap rolls is larger than the rest of them; and all of them should be graduated from back to front, the front being the smallest. This gives the back harness more sweep than the front one; and starting with the back roll and making each one a little smaller than the other, secures that the yarn is lifted to an equal height, thus giving an even shed. Make each of the ends of the rollers alike. It will not do to have the size of the rolls different at each end, as by this the sweep is altered, and in trying to get the shed right at one end the other end will be made all wrong. It is not necessary for the hanging of the harness in a roller or cam loom to be a difficult job, and yet there are many who have to do a great deal of changing and trying before they get a proper shed. I find it a great advantage to arrange the underneath straps, so that in unhooking them to take the harness out the length is not changed. When so arranged, all that is needed in starting a warp will be to hook them up, and if the shed was right before it will be right now.

For plain work, where the same is continued year in and year out, these plain looms have much to recommend them. The weavers can handle them in any way they choose, and the shed is not changed, and the high rate of speed at which they can be run gives them a great capacity for a large production.

The side-lever loom.—The difficulty in making changes in the various forms of satinet weaves on roller-looms makes the side-lever loom a very desirable one where a variety of plain weaves is used. The style of side-lever on the excellent loom known as the Gilbert loom, may be found also on others, with slight alterations. The cams consist of thin plates having grooves on one side of them; for different weaves the grooves are made different, having a longer or shorter dwell as may be required by the weave.

The jacks or levers have a stud cast on the

side which fits in the groove of the cam-plate. On some looms a roll is fitted over the stud which makes it easier for it to follow the groove.

The cams are held in their proper "time" by a long feather or key on the cam-shaft, and the position in which the cams may be set depends entirely on the cutting of the key-slot in them. A four-harness twill would have the cams numbered from 1 to 4. These, put on in that way, give the twill correctly. If you wished to reverse the twill the cams would be put on beginning at number 4.

To break the twill it would only be necessary to put them on as you would break a twill in writing, as 1—3—2—4.

Gears.—Different gears have to be used for different weaves. I might give some numbers for the Gilbert loom, and also for the Stafford, the Davis & Furber, and others, but to make it complete would take up too much space, besides being unnecessary. One can easily study out the gears he wants if he has no basis from which he can figure. There are a great many places where no record is kept of such things, and one can only get the right thing by trying. A plain cotton weave would be geared so as to make one revolution of the cam-shaft for every two picks, providing there is but one lift to the cam. If it raises the harness twice for every revolution it makes it has two lifts, and consequently it only has to run one-half as fast, or one revolution for every four picks. A four-harness cassimere twill will make one revolution for every four picks. A five-harness docskin has to make one revolution every five picks.

These side-lever looms, when the weave is changed, and every thing is in proper time, need less altering and fixing than any other kind of looms, and where plain work is made they are a very satisfactory loom to all concerned.

CHAPTER V.

THE BOX-MOTION.

The old cam-motion.—We will now take up the box-motion, beginning with the old style cam-motion, that which has been the terror of loom-fixers where a number of shuttles had to be used and perhaps a peculiar shaped cam. The motion should be perfectly steady on this, but I have frequently found the bearings of the tappet-shaft worn very much. This is gross carelessness on somebody's part, and it is more aggravating when we consider the importance of having everything connected with the box-motion in perfect order.

If the bearings are worn, get new ones, and have them fit on the shaft, so that there is no play in them. In putting the shaft in, take the trouble to have it perfectly true. There is no room for anything but absolutely perfect work in dealing with this thing. Let it turn easily in its bearings, not loosely, but a " fit."

Ratchet-wheel plate.—On a double reverse motion the first plate for the ratchet-wheels is keyed on. This should be well done, so as to avoid any twisting of the ratchet plate after you have it set. The key should be the full width of the key-way, not one that wedges in and cracks the hub of the plate. When this plate is fastened on, put the ratchet wheels on to it as they belong. Do not make them very fast until they are set.

The forks.—On the tappet plate put a pair of ratchet-wheels; then get the forks and lever, or hanger, ready. The large stud in the lever will, if the least worn, allow the lever to sway to and fro in working, and often it will sway enough to make the fork miss a tooth on the ratchet. This should be fixed, if too badly worn. If the forks are old and worn blunt they should be filed. This job should be nicely done, and by care the point can be shaped to draw into the ratchet. They can just as easily be filed so that the point will slip out, so it pays to be careful.

I would not use a fork if required for reverse, if it had to be filed too much to bring it right. They are of course shortened, and the points are wider apart than they should be, making trouble in getting the chain to vibrate the fingers enough. In filing, be sure that the top ones are alike, and the bottom ones the same. You cannot fully determine just how the forks are going to fit until they are tried on.

The stud on the top of the lever should be raised high enough to bring the pin that holds the fork in the lever in line with the tappet-shaft. This will usually be found to be the top of the slot. Now put the cam-plate on the bottom shaft, having previously fitted two cam-blocks on it, each exactly alike in regard to their shape, and also as to the distance from the centre of the plate. This insures an equal push by each cam-block. The cam-plate should now be turned so that the blocks push the lever to its farthest extent, and then you can set the ratchet wheels against the fork.

Setting ratchet and plate.—The object sought in setting the ratchet with the forks pushed to their fullest extent is to get our tappet-plate fastened where it ought to be, and there is no other way that can be employed in doing this, unless you guess at it, as we shall see. Now, if the fork is pushed as far as the blocks are set to push it, we can set the plate and the ratchets, both forward and reverse, and we know they cannot be pushed any farther. We want the plate to be set so that the centre of the truck-roll stud is directly in line perpendicularly with the centre of the tappet-shaft.

We put on a tappet to guide us in setting, and turn the plate by hand to its right place. If the friction band is on you can tighten it so as to hold the plate where you put it, having your hands free to set the ratchet-wheels.

Set the top one so that you can just lift the fork out of the tooth of the ratchet-wheel. Set the reverse underneath in the same way. You will find, perhaps, that it is hard to move one so as to set it without throwing the other out of place. But it should not be neglected in the least, and if, after trying it when everything is fastened and ready to start, you find it is not just right, do it over again. When it is set right, the fork should be so that there is hardly room to lift it out of the tooth of the ratchet when the plate is held tight by the friction-band. Lift it out of the top or push side, and press the bottom fork in the reverse side, and if they go in with a snap they are right, providing your plate is where it should be. Getting the plate right, and at the same time setting the ratchet-wheel is the prime object.

The friction band.—The friction band is of far more importance in the successful running of the box-motion than most fixers seem to be aware of, judging from the careless way in which some of the bands are made and kept in order. I know of nothing that gives such good returns for the labor expended in doing a good job than on this. The band should be made out of a piece of iron of good quality, so that you can make holes for the rivets and not have the band break at the holes. Get the band as wide as possible, but be very careful not to get it too wide, as some cams or tappets fit the plate differently than others, and would be a source of trouble if they should cause the band to bind. Line the inside of the band with a piece of new leather; it should be the best you can get, and uniform. Put it on as tight as possible, and cut the leather so as to let the rivet-heads sink into it, and not come out flush with the rest of the leather.

Effect of a good band.—The effect of a good friction-band is to give a smooth, easy movement to the boxes. If the leather becomes glazed, the friction lets the plate turn in a jerking manner, and the same motion is imparted to the boxes. An important advantage in using a good leather lining is that it takes less friction to keep the boxes from dropping, consequently there is less strain on all parts.

Timing the starting of the boxes.—In timing the starting of the boxes there is little room for variation. The tappet should move just enough to raise or lower the box $\frac{1}{8}$ of an inch when the protector strikes. On some patterns where you skip a box, or perhaps two, it is sometimes necessary to start it sooner; but there is risk in turning it too soon, as the swell on the next box is liable to open out the protector and let the loom keep right on running with a shuttle in the shed.

Regulating the straps.—In regulating the straps that connect the filling chain fingers with the fork, let the lathe be back; this allows the fork to go back as far as it will go. Adjust the straps on a reverse ball first. Do not fall into the common error of having the connection too short. It is done to prevent the fork from slipping out of the ratchet. If the fork is inclined to slip out, the strap cannot hold it in, so there is no excuse for having it too short on that score.

Set the chain on a reverse ball, and adjust the strap so that it draws the forks up into the ratchets underneath. They need only to touch the ratchet wheel loosely. If too tight the straps draw on the fingers, and when the chain is turning from a blank to a reverse bar the pressure will occasionally slip the chain on the cylinder. This will sometimes confuse the fixer, and thinking that the forks missed catching in reversing, he tightens the straps and increases the trouble. If made so that they catch, it is enough, and it leaves room for the forks to drop into the top ratchets. There will be no trouble in missing if the points of the forks and the tooth of the ratchet are of the right bevel.

Worn fingers.—Sometimes the under side of the fingers are worn at the place where the ball rests. If so, they should be filed so that the finger is straight. If not, it lifts with an unsteady motion which will shake the forks and make them liable to miss. To remedy this, weights are tied on the forks, and I consider these entirely unnecessary. If the fingers are made to work smooth and easy, and the straps properly adjusted, the forks are heavy enough. The filling chain should not be set to move until the forks begin to retreat. If moved too soon, the fork is brought in contact with the under side of the ratchet-wheel on reverse motion, and will slip the chain the same as with a tight strap.

Friction of the chain-cylinder.—The friction of the chain-cylinder is a particular thing to adjust. Some fixers will not allow the weavers to oil the stud on which the chain-cylinder runs. If this is not done the back of it becomes dry and the cylinder turns with a jerk, and sometimes will not stop when it should. The stud should be oiled, but not the friction end of it. The leather washer used for friction should be made of new leather, and the inside cut out with a sharp tool, so that it fits very tight on the square shank of the stud. It should not be too thick, as the iron washer should also fit on the square shank to keep the whole from turning with the chain-cylinder. A little care is needed in making this friction what it should be, but it is a part that can put the fixer to lots of trouble if it be not right.

Connecting the rod for turning the cylinder.—There are various ways of connecting the rod for turning the cylinder on the pump-motion. The best way is to turn it from the eccentric plate, on this style of loom, but many are run from the pump-rod. The objection to the latter way is, that the unsteady motion caused by the parts being worn is imparted to the motion of the cylinder, and this should be avoided whenever possible.

The filling chain on the other styles of looms, excepting the 1880 and later kinds are turned by the pick and pick gear in connection with cam-block plate. This gives no trouble, and is so simple as to require no particular mention.

The box-motion.—The box-motion on the 1880 loom with the improvements up to the present time make this as near perfection as one could well wish. We hear complaints from some, but that will always be. It will require skill and good judgment to run any loom that is capable of producing the variety of goods that the markets now require, and which will increase in their complication from year to year.

On the 1880 loom, the box-motion is composed of a series of levers, and their operation, in connection with the complicated arrangement of gears, offers a very uninviting field for careless fixing. If kept in proper order, I think this motion is more reliable than the old tappet-motion in avoiding changing or missing; but the arrangement, taken as a whole, calls for sound mechanical ideas on the part of the fixer. It can soon be ruined by botch work. I like the idea that we have a chance to exercise our skill in doing a job in a scientific manner on this box-motion.

Putting on cylinder-gears.—In setting the parts of this, we will begin by putting on the cylinder or quill-gears, the vibrating levers being already in place. We will put the gears on without setting them in any particular place until we find things in their proper positions. The cam-gear stud may need to be set, and to find that out, we put the gear on, and then try the "dwell" cylinder-gear. The dwell should fit almost tight on the slide made for it on the cam-gear; if it does not the teeth of the cam-gear will fail to mesh with those of the gear on the bottom shaft, and you know what the result is.

I have often known these teeth to catch, and fixers were at a loss to comprehend the cause. There is no other cause for this mishap than the dwell on the cylinder-gear being too loose, and thus failing to guide the teeth in their proper place. To set it up tight, observe that the stud for the cam-gear is turned on an eccentric. You can loosen the set-screws and turn the stud wherever you wish.

In setting the gears you will, after putting on the cam-gear, notice that there is but one place that the dwell-gear can be set, so you cannot get that in the wrong place. The other cylinder-gear has to be set, and if not in the right place to a tooth it will make trouble.

Setting the cylinder-gear.—To set it is very simple. Slide it on the stud and press one of the vibrator levers as far as it will go towards it. Set the gear, so that when meshed with the intermediate. and turned over and towards the lever-gear, to take up the play in the teeth of them, the last tooth of the cylinder-gear barely touches the teeth of the lever-gear. This allows the latter gear to get into its place before the cylinder-gear starts to move.

Sometimes you can find many of the cylinder-gears worn for four or five teeth. When they are in this condition the box makes the attempt to get to its place, but there not being enough teeth in the cylinder it cannot turn the lever-gear far enough. The reason the cylinder-gear is worn is, that the lever-gear has been gearing into it after it began to turn. The fault lies in the gearing or timing of the cylinder-gear. It starts so soon that the lever-gear strikes it while going at full speed, and thus begins to wear them both. If the cylinder is set as I have described the lever-gear is well into place before they start.

Boxes Missing.—This may also explain why we find the box on one side dropped down, while the other side is all right. The gear which is working pretty close happens to miss one of the lever-gears, because the teeth on it are worn, or perhaps the lever-gear does not gear into the cylinders far enough. That is the place to look for the cause of a box missing like that. There is some little thing about the gearing of the lever that is wrong. I mention this because I have seen the filling chain, and the fingers on the filling chain tinkered with when the whole trouble lay in the levers and gearing.

Levers.—Now, without waiting to be driven to it by the loom refusing to work right, suppose we just look over a set of levers, and see how they are and how they should be. If the levers have ever been filed, either on the top or bottom, or the vibrating finger, the only way to put them in good shape is to put a new one on. There is never any room for too much metal about the levers. They could not be put on and tried each way with the indicating fingers without binding, and I always found these box-motion levers well handled when they left the shops. The only time they get filed is when the stud for the indicating finger lever gets set wrong and binds the vibrating finger too tight one way, or something like that occurs which only requires a slight alteration. But, without counting the cost or looking on the other side of the queston, the fixer files something. The metal once removed cannot be replaced again, although I have seen it tried. Yes, indeed, I have seen such botch work on these looms as required pins to be inserted in the tops of the levers to compensate for what had been filed off. A pretty mess! It is always safe to let the filing alone until you have become thoroughly satisfied that you are right; and in regard to these levers I am very certain that if one tried very hard, he could be easily satisfied that they needed no filing. The grates at the bottom of the lever are the only things that I see about these looms, or rather about this box-motion, which in any way needs filing. In fact you can use this grate sometimes to help you in making the vibrator fingers and the indicating fingers fit.

Easing the Levers.—If, in trying the indicator fingers into the vibrators, you find they bind, try the other side and divide up this undue strain on the levers by moving the indicator lever stud. This done, you can file each side of the grate-slot, and thus ease up the levers when the indicator raises up into the vibrators; and at the same time, by taking it off each end of the slot, you cause the lever-gears to gear deeper in each of the cylinders instead of only one of them. Aim to keep the vibrator fingers in such shape that when the indicator presses into them it holds the lever-gears firmly into the cylinder; there should be no play or rattle about them at all.

Fingers for the vibrators.—There are two kinds of fingers to operate the vibrators. The first kind worked in connection with the underneath fingers. There was considerable dissatisfaction with them, and in some mills they were taken off and replaced by the fingers which acted directly on the vibrators. My experience with the first kind of fingers has taught me that the principal trouble is that the fingers are not heavy enough for the improved vibrator finger. It will give trouble in other ways, but I think it is easy to keep it in proper shape, so that it will run right along.

Running the filling chain.—There is a point in running the filling chain, which may well be heeded. The underneath fingers are curved and their points reach considerably further than the place where the ball rests when the chain is stopped. The ball, in coming up, begins to lift it early, and, being at the longest point of the finger or lever, lifts more slowly and consequently more steadily, and by the time the chain stops turning, it has lifted the long finger into place with an easy motion. Let the chain turn from the loom, and the ball comes under the finger at the short end of it, regarding it as a lever, and it lifts it suddenly and quickly into place. Now, this jerking the lever up gives it a bound which it cannot recover from in time, and it is liable to cause some of the vibrators to change or to do harm in some way. A moment's study of the principle of this combination of levers will suffice to show how natural the movement is; but some men never think, and that was the trouble with the fixer who considered that anything was good enough, and so some of his cylinders were turned from the loom and some towards it. He had trouble with the fingers; they would fly up and come down with several bounds. Hooks were inserted in the bottom of the fingers, and springs applied to each of them. For some reason, he could not tell what, the points were broken off the underneath fingers and the end rounded to a nice blunt shape. Then, when he did run the chain the right way, the fingers were so blunt that when the balls struck them they were lifted with such force as almost to throw the long lever clear over; and so things went, until in about a year from the time the looms were put in, new from the shop, there was very little left of the original box-motion. This is the result of reckless altering without considering the relation which the thing you are working on has to something else.

Other considerations.—More could be said about this box-motion. I might describe the adventures of the fixer, who, for a whole day, worked on a loom to remedy the effects of changing the star-wheel—a thing that with ordinary judgment could be pulled to pieces and replaced in five minutes. When you alter one thing without first thoroughly understanding what you are doing, you cannot expect to fix it by altering something else.

The fixer who is successful on one kind of loom is successful on all kinds, because he thinks. A new style of head or box-motion has no terror for him, because he reasons that there is a principle about machinery, and the makers of a loom cannot depart from it and he can follow out their idea from a mechanical stand-point and get all the good there is in it.

The fancy box-motion.—The fancy box-motion made by Crompton, to apply on looms of the old tappet style, is a very excellent arrangement indeed, and where mills are on work which requires a good many shuttles I call it an invaluable improvement. The arrangement is practically the same as on the new loom, being altered only to suit the difference in the loom. The filling chain is turned with a hook as on the old style. It is not perfect, but such as it is, it is a great stride in advance of the old tappet motion.

The 1883 loom differs in that the levers for the box-motion are placed in a horizontal position, thus doing away with the upright levers or vibrator levers. The balls on the filling chain lift the fingers that ride on the chain, and they, in turn, lift the levers below on which are the lifting or vibrator-gears. The two quills are placed one above the other, between which the vibrator-gears are placed. The action is more direct and it does away with some of the difficulties experienced with the earlier kinds.

CHAPTER VI.

STARTING A WARP.

GETTING A WARP READY—CONDITION OF THE HARNESSES—
REPAIRING HARNESSES—HOOKS—NUTS ON THE HEDDLE—
WIRES—RULE FOR THE NUMBER OF HEDDLES—DRAFT-
DRAWING IN THE WARP—THE RIGHT KINDS OF REEDS—
IMPERFECT REEDS—REEDING THE WARP—FINDING THE
WIDTH—MEASURING—LEASE RODS.

Repairing and starting a warp.—In the preceding chapters the principle subjects have been touched. The minor details have been omitted for want of space. We will proceed to get a warp ready and to start it, and will speak of the various things that come in our way as we go along.

Drawing in the warp.—The drawing in of warps is a subject that seems to have different degrees of importance attached to it in different mills. Of course the drawing in can only be done in one way; but the work incident thereto can be done in several ways, and in most mills there is a chance for improved methods to be worked in.

Condition of the harnesses.—I think it is a fact that in nearly all mills of ordinary size the care of the harnesses devolves upon persons who think their duties are not anything more than merely to clean the harness and reeds. No attention whatever is paid to their condition. Sometimes a hook has broken off in the wood underneath while in the loom. The fixer, unable to get at it to make a proper job of it, has tied the wire on with a piece of lace leather. When that harness once more gets to a loom, the lace has to be resorted to again. The same with hooks for the heddle wire. Then, again, the pins that fasten the heddle wire get lost. The harness goes in the loom all the same and the bar or heddle wire, being loose, gets kinked and causes the top or twisted ends of the heddles to catch on the next harness and they are spoiled. So is the wood or frame on which they caught. This is a picture of no rare occurrence, but rather of a very common one. The harnesses are a much abused article in my opinion and it is because no one seems to attach the importance to them that they deserve. If other things were to be neglected as they are, we should think the weave room was going to ruin.

Repairing harness.—I would have some one responsible for the care of them who was able to keep them in good repair. A set of harness nearly always needs some little repairing after weaving a warp out, and the better they are cared for, the less accidents will happen to them. The hooks should always be uniform. In some mills where there are different kinds of looms, different lengths of underneath wires are used, and even top-hooks have to be moved to suit the different looms in which the harness goes. This should not be. The distance apart on all the harness in the mill should be of one standard. The moving of hooks ruins the harness-frames by wearing out the holes, and in trying to fix the hooks so they will not come out the harness get split.

Hooks.—The inside hooks should be put in so that they are all of the same length from the inside of the frame. If not, one hook bears the strain of the heddle-wire more than the other. Care should be used that the hooks are not screwed in too far. Sometimes they are put in so far that the top or twisted end of the heddle touches the frame when you try to unhook it. This should always be avoided, for it causes the heddles to get bent while the bar is being unhooked to move them.

Center-wires.—The center-wires are a good thing and there is no reason why they should not be on all harness, and a great many reasons why they should. The heddles will last longer, the weaver can move the heddles more freely in putting in threads; they save the harness and help the fixer in determining what strain he is putting on when hooking up the harness.

Nuts on the heddle-wire.—The easiest thing to neglect is the nuts on the heddle-wire. They are so small that the thread is poor at best, but being out of the fixer's line of business, they do not get tightened up when in the loom; and when in the drawing-in-frame, they give no trouble and are neglected there till the heddles are spoiled. There is no way to keep such things right but to look for these faults. A set should be carefully examined before the warp is drawn in and everything put in order.

Threads on the butt.—Provide a die to re-cut the threads on the butt. It will be some trouble, of course, but the trouble caused by bad harness in a loom, both to the loom-fixer and to the weaver, is a hundred times greater than the work required to keep harness in proper order, and far less satisfactory. A loom running on nice work, with a good many harness, is at a great disadvantage if it has a poor set of harness in it. It will profit the overseer, as well as all concerned, if the harness receive attention and are kept in good repair.

The size of heddle.—The size of heddle to use is a subject on which opinions differ. My experience has been that it is not best to use a finer number than 24, and I like a 23 the best for all purposes. On Blankets and such goods, that take yarn of a coarse quality, I would have number 22 heddles. Some overseers want a very fine heddle when finer yarns are used, and when the warps contain a large number of threads. They give as a reason, that the large heddle crowds the warp. Well if a large number of threads should be drawn on few harness, that might be true, but such is seldom the case in fancy cassimere mills; and, if true, it is unnecessary. I would not draw that way if it crowded the warp, but would repeat the draft. We want a heddle that will give the best results, all things considered. A thread will be less liable to meet with friction, in the intersections in the eye of the heddle, in a coarse one, than it will in a fine one. Then, there is the advantage of a stronger and better heddle lessening the chances of heddle-smashing. I have used them on fine worsteds when over 6.000 threads were employed, and reeded only 66 inches wide, and I found the results very satisfactory. However, it will be a profitable and satisfactory thing for overseers to make a trial of this thing, and thus see for themselves which gives the best results.

Rule for number of heddles.—To estimate the number of heddles required for warps with a cross-draw, I give the following simple and natural rûle. Multiply the number of times the thread is drawn in on each particular harness, by the number of patterns in the warp. Thus 5408 threads, 26 in pattern, 208 patterns in warp.

DRAFT.

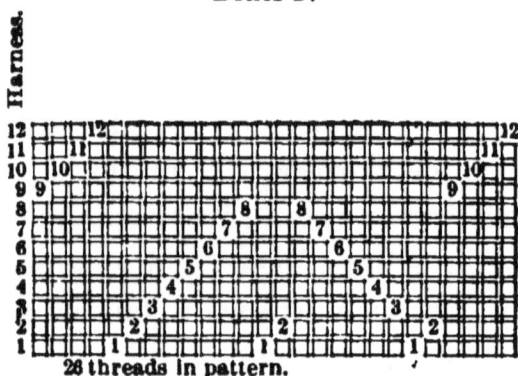

26 threads in pattern.

Number 1 is drawn 3 times 3+208=624
   "   2 "   "   3 "   3+208=624
   "   3 "   "   2 "   2+208=416
   "   4 "   "   2 "   2+208=416
   "   5 "   "   2 "   2+208=416
   "   6 "   "   2 "   2+208=416
   "   7 "   "   2 "   2+208=416
   "   8 "   "   2 "   2+208=416
   "   9 "   "   2 "   2+208=416
   "   10 "   "   2 "   2+208=416
   "   11 "   "   2 "   2+208=416
   "   12 "   "   2 "   2+208=416

Number of heddles, . . . 5408

Drawing in the warp.—We are now ready to draw in the warp. In this work the overseer

78

may also interest himself more than is usual, and do good. The work is mostly done by younger persons, or by those who are not in a position to judge fully the importance of the task they are performing. For their good, and for the benefit of the loom-fixers and weavers, it is best for the overseer to see to it that the drawing-in is done after his own ideas. The drawer-in should be provided with suitable hooks for both drawing-in and reeding. The time lost in working with poor hooks is quite an item. By having a system of measuring, the heddle bars can all be hooked up while in the drawing-in-frame. They should never be taken out until they are. The damage done by leaving the bars unhooked is unnecessary and very aggravating to the loom-fixer, who is compelled to put in the threads broken by the tops of the heddles, sometimes in whole bunches.

The drawer-in should be held to strict account for mistakes made. Frequently the errors made in drawing-in are taken as a matter of course. I find that there are less of them if the drawer-in is made to bear a share of the responsibility.

The right kind of reeds.—The reeds used in the weave-rooms of many first-class mills are far from being what they should be. They cannot be handled too carefully. A reed should never be thrown on the floor, or laid on any but a flat surface. It is an easy matter to kink or bend a reed, and once bent it is like a saw-blade, it can never be perfectly straightened.

Bending reeds.—The practice of bending a reed in the loom with the hammer, alluded to in a previous chapter, should not be permitted. This causes great damage to the reeds, and makes trouble for every one. When the dents are spread apart, or bent, great care should be exercised in repairing the defect. It is sometimes done with the blade of a knife or the point of a screw driver, a very clumsy way, as any one will admit. A pair of reed plyers should be used, and the reed should be made as near perfect as the eye can judge.

There are many kinds of fine goods that are seriously damaged by wide splits. In former times a slight streak in the cloth would not be noticed; but nothing short of perfection seems to suit now.

Imperfect reeds.—Among the ill effects of an imperfect reed perhaps none are more important than the confusion of the weavers by wide dents. They get accustomed to seeing the streaks in the cloth as they are weaving and very often a wrong draw goes unnoticed under the impression that it is a wide dent. The cloth should present an even and perfect appearance to the weaver whose attention is then easily attracted by every imperfection, however slight.

Poor reeds are expensive.—The quality of our reeds should be carefully noted. A poor reed is a very expensive thing. We often find a reed that will bend easily. If the dents are soft enough to bend readily they are usually soft enough to wear, and the effect of a worn reed on a fine warp is well known. If the overseer is careful to notice how the reeds obtained from different makers turn out, as to quality, he can soon determine how to keep the room supplied with the best that can be had. The flexible bevel-dent-reed lately introduced commends itself to overseers for the many good qualities it possesses. The dents when spread readily spring back to their place, and the dents being beveled are a great help to the warp. I regard it as an invaluable improvement over the old style reeds.

Reeding the warp.—We will now turn our attention to reeding the warp. To make calculations for reeds, determine the threads in a dent and the width you want to make it:

Example.—3600 threads in warp, 4 threads in dent, must be 72 inches wide. What is the reed?

$$4)\overline{3600}$$
$$72)\ \overline{900}(12\tfrac{1}{2}$$
$$\underline{72}$$
$$180$$
$$\underline{144}$$
$$\overline{36}$$

12½ reeds is what you want.

Another case: 5600 threads, 6 in dent, 14 reed. This will give the width:

Example: 14 reed
6 in dent
threads per inch 84)5600(66⅔
504
560
504
56

Finding the width.—To obtain the width, the threads and reed being known and the reeding being irregular, the only thing to do is to get the average, as near as possible, of the number of threads per dent.

Example: 5408 threads, 15¾ reed reeded 5—4—4—4—5. There are 6 dents in this reed pattern, and the number of threads contained in those 6 dents you will see is 26. Dividing by the number of dents will show that there are 4⅓ threads in each dent. Now multiply the reed, 15¾ by the dents per inch as in an ordinary case; 15¾ x 4⅓ gives us 68¼ threads, divide the threads in the warp by 68¼, and we have 79¼ inches nearly.

There are rules given which will work in certain cases, but you can see your way clear by this method, and if you invent your own special rules you will be more apt to remember them. The principle which is here given is very plain.

Measuring.—The drawer-in, having been given the width which the warp is to be reeded, should measure off accurately so as to get the warp in the center of the reed. If a mistake in measuring is made the fixer has not room on the ends of the reed to slide it, so as to bring the warp in the center; and to accomplish this the reed is cut off at one end. It certainly should never be done. Why spoil a reed to get around a mistake temporarily? If the warp cannot be brought in the right place in the loom, it is far better to reed it over again. There is little excuse for having a warp reeded so far out of the way, however, and it does not often occur.

Lease-rods.—The lease-rods should always be left in, even though the selvage may be drawn in on the regular harness. You cannot tell when you may need them.

CHAPTER VII.

WARP-MATTERS.

Avoiding damage.—In starting the warp, a great many of the damages to the goods, that are caused by some part of the loom not being properly adjusted, can be prevented by making sure that each part is in proper order. There is no weave-room but that has mistakes made in it every day, and always will have; but there are lots of these mistakes that begin when the warp began, and we can keep them down by keeping a sharp look out in starting the warp.

Lifting in the warp.—To hold the harness up when we lift the warp in we need a stick or rest. I have seen a hook hung over the top of the frame of the loom, but I do not think it as good as the rests made of wood. They should be hollowed on the under side to make them light, and should be high enough to hold the harness within one inch of the height they should be when hooked up. This leaves the bottom wire slack enough for you to hook on before hooking the top, which you will do on all looms excepting the pump-motion.

Hooking up the harness.—If they are hooked on top first, the straps on the bottom are too tight to permit of the wires being hooked, unless you let the harness down. In hooking up the harness, be sure that the head-motion is closed. You can then level up the harness which should be carefully done.

Drawing in the selvages.—After we have the harness hung, we will draw in the selvage. The straps and heddles usually employed are none too good. In very many cases they are utterly unfit for the purpose. The idea is quite prevalent that anything will do for the selvage. It will strike some people, after a while, that the selvage is a very important part of the cloth. It is a common thing for most of the selvage threads to be thrown back in the weaving, and they are never put in again during the whole warp. Sometimes they cannot be put in for want of heddles. One strap may contain six or eight heddles, and the other side two. Sometimes eight or more double threads may be drawn all right on one strap; while on the other, for want of heddles, four or six are put in each heddle.

In some mills the character of the selvage is kept uniform on all of the goods. It is dressed in three colors, four threads of each. This gives a neat appearance to the goods, and rightly gives the impression that everything pertaining to them receives the utmost care.

A good selvage also protects the warp in weaving. It enables the temples, or temple-hooks, to get hold of something that can stand the strain put upon them. It should be of uniform width on every piece woven—one inch in width at least.

The beam friction.—Having drawn in the selvage we next put on the beam friction. A good substantial friction should be used, and care should be taken in its construction. Uneven cloth causes more damage than any other fault in weaving; and it is being produced every day in any weave-room. The trouble resulting from it is incalculable. In most cases uneven weaving is almost imperceptible. It is not always detected even on the perch, but in the finishing-room it begins to show up in the shape of shaded goods, cockles and the like. It confuses the finisher, and sometimes makes a bad matter worse; so we cannot be too careful about everything connected with the beam.

Ordinarily a heading is tied around the beam-head. It is a good thing, and if nothing else is used it should always be put on. There are other things that can be used also.

Difficulties with beams.—The chief trouble with beams letting off unevenly may be found in the way the chair holds them. It can easily be moved, and if it slips one way or the other, to cause the shafts to bind against the frame of the loom it will not let off easy. The chair should have a solid foundation; not an uneven packing of leather, but something on which it can rest square and solid. It should be bolted very tight, so that the weights do not pull it over and thus cause the shafts to bind.

Unless the work is extremely heavy, I should always set the chairs so that the shafts of the beam can be seen to have room to spring to and fro as the pick is beat in. It may be just perceptible but it will answer the purpose. If the shafts are tight the beam binds. On very light work the beam, if resting on the chairs, will take so little weight as to be hard, or may be, impossible, to regulate. In that case provide blocks to rest on the girth, and long enough for the beam shafts to rest on them, and be held off the chair.

Putting leather in the bearing for the beam on the frame of the loom will lift it up; but blocks are better, for they do not hold the shafts so rigid, but allow them to work back and forth;

and this action seems to let the beam–head work around, and not slip by jerks.

Width of band.—The band should not be too wide. They are often made wide to give them strength; but sometimes this goes too far, and they are made so wide that they bind on the sides of the groove. If cloth is not tied around the head, it is equally good, and in some cases better, to let the head rest on the bare chair and then put a clean thrum under the band. This the weaver should change at the beginning of each cut. This will insure against having the beam running with a glazed friction. It looks like a good deal of work, but I can assure the reader that it is a profitable thing to do, both for the sake of the uniformity of the weaving and for the help it gives the warp, if a tender one, by letting off so much easier, and uniformly each pick.

Tying in the warp.—In tying in the warp the ends should be combed and well straightened before tying. If a thread here and there is slack, you are not sure of them all showing up when you look the harness over. Sometimes a wrong draw may be missed in consequence and you cannot afford to take any chances in this work.

Care in tying in the warp is time saved, for many threads may be broken out at the start by the shuttles if the shed does not open out free.

Fastening the reed.—In fastening the reed, measure each side so that you are sure to get the yarn in the center. The blind nuts in the slat that fastens the reed are, in many cases, spoiled, and a nut is placed on the outside. Each time the reed is taken out these have to be laid on the breast beam while the process of tying in, etc., is being gone through with. When you want them you may find them where you put them, or you may find them under the loom,—a great annoyance always. Now, why not take the slat out and get the nuts fixed and put where they belong; it is such neglect as this, in all the simple things about a loom, that runs a section down and keeps the fixer on the go from morning until night.

The reed should not be screwed perfectly tight with the slat before the cap is put on. It does not allow it to adjust itself and the cap when put on may spring the reed. The cap should never be pounded hard. This bends the dents of the reed and is a damage to it.

Handling the chain.—We now have the warp ready for the weaving. Let us look at the chain. On old chain, to make perfectly sure that there is no danger of mispicks being caused by it, it is best to turn it by hand all the way around and look at and feel each bar to see if it is possible for the risers to slide one way far enough for a jack to slip off. Do not tie up a chain at random and let it go. If you tie string between the sinkers on a chain of 16 harness, you do it on the part nearest the back of it, so that there are 16 or 18 spaces, each of them worn slightly on the ends that come together. Being crowded tight, it makes them worse than they would be if the bar were not tied at all. If you tie your chain so that the ball nearest the string comes under the jack-finger, or if the jacks are moved to one side to make them come over a chain so tied, either one side or the other will not come right. It will take some manœuvring to make them come right.

The loom should not be started until the chain is fixed so that it is impossible for a jack to slip off the balls.

Putting on the links.—The links of the chain should be put on all alike. A chain is always liable to make mispicks if the links are not put on right. The link can catch on the end of the cylinder and drop again, just as the shed opens, so you cannot see what caused the mispicks. If you are troubled by having mispicks occur once in a while, but not often enough to enable you to watch the loom and see what did it, just try the plan of doing the following simple things in a thorough manner and you will not be troubled so much.

Points about links.—Links should be put

on the chain so that the outside link on one side of the chain is opposite the outside one on the other, the inside link opposite an inside one. This causes the chain to hang plumb and keeps it straight while passing over the cylinder. Another way of putting links on is to let them overlap each other like shingles on a roof. The end of the link pointing towards the cylinder should be the outside end and the other end of the same link coming on the inside of the next one precludes the possibility of the end of the link catching in the groove on the cylinder. This way of putting them on shortens them slightly and is not so well on new cylinders. Whatever way is employed, be sure to keep both sides alike and directly opposite each other.

Building filling chain.—In building the filling chain the same care in regard to the links is necessary. Trouble with pattern changing can be averted in one-half of the cases, if the chain is built right. On reverse-motion, the reverse-balls should be callipered. They have to be exactly alike for reverse and you may as well take a little trouble here, as to spend your time fixing the loom.

The pins of the bars.—The pins for the bars are another place where work is often slighted. They should be of uniform length and not too large. If a pin feels as if it is nearly broken when you bend them to make them stay in, take it out. A chain coming apart while running is likely to do a great deal of damage.

The size of the shed.—We are now ready to weave the heading in, and while doing so let us regulate the size of the shed, and also its height. Do not make the shed too large. It is common for fixers to run with the shed opening the full width of the reed. The aim of a good fixer is to make the loom run without making it too large. If the shuttles do not run straight a large shed will not help them. You are only trying to get the yarn out of the way of the shuttle, instead of keeping the shuttle where it belongs. Try your shed in various ways to make sure you have everything right about it. Be very careful not to let it bottom too hard. It should just clear the race-board when the shed is opened full width. The harness should be leveled so that when the shed is opened the top-threads of it are even. If one harness is permitted to work lower than others, the shuttle will occasionally slip over a thread. This can be avoided by keeping them even.

On warps with a backing on them, you sometimes have to lower some of the backing harness to get them down on the back-pick. In such a case, you cannot help having that harness lower when it is raised, but it will not matter so much on a backing harness.

If you have difficulty in getting the harness low enough, you can help it by making the underneath connections extra tight. Do not leave the loom until you are sure that the harness is working so that it makes a good shed. It may be working right while you are watching it, and after a time the connections get slacker, or may be the warp does not run as tight on the selvages as it did while you were at the loom, and if any of these things happen you may have a cut full of harness-skips on the back. So, to be on the safe side, see to it that you have a shed adjusted in such a way as to meet any contingency that may arise.

Examining the harness.—We now have to raise the harness one by one, to look them over. Inasmuch as this operation is the only safe one that we can have to make sure that there are no wrong draws in the warp, it is important that it be very carefully done. A few minute's time is of little consequence compared with the time it will take to fix a wrong draw if one is passed. The safest way is to have a white rod to put under the threads of each harness as they are lifted. The run of the pattern should be followed the entire length of each harness; and do not trust to the eye too much.

On some harness the pattern may come up with a colored thread at regular intervals among black. In such a case, should a colored thread be left out and a black one come up in its place, or the colored threads come too close together, the eye can readily detect the difference; but where there is a pattern on each harness, there is no safe way but to count the patterns over as you go along. On mixes it is sometimes very difficult to decide which is the right thread; but there is no other way than to examine closely. It is sometimes necessary to use a glass. On silk and black, double and twist, if the silk is very fine, the only way to look them over is to take each thread and untwist it. This may require hours to do, but you cannot get out of it. It will not do to risk anything on such goods.

It is also a good plan to weave a piece and then to make another heading. The piece need not be more than three inches long; and when the lap is woven down, take the piece and have it scoured. When dry, wrong draws will usually show up plain enough to be seen. None of these precautions should be neglected upon any account.

CHAPTER VIII

SHUTTLES, TEMPLES AND BELTS.

GOOD AND BAD SHUTTLES—SHELLACING SHUTTLES—POINTS OF SHUTTLES MEETING IN THE SHED—INJURY DONE BY SHUTTLES—TEMPLES—THE USE OF HOOKS—STRAPS—THE DUTCHER AND OTHER TEMPLES—SETTING THE TEMPLE—METHOD OF PUTTING ON TEMPLES—THE BELT ON THE LOOM—OILING THE BELT—PUTTING ON THE BELT—LACING—BELT-SLIPPING.

Good and bad shuttles.—In starting the pattern, before putting in the shuttles, we may as well look them over and see if they are in good condition. I do not think it an exaggeration to say that fully one half of the shuttles in use are unfit for the purpose; and in most cases their condition is due to neglect. We find them worn flat on the tops and sides. If anyone says that this cannot be helped, I would ask him why it is that on some looms, and with some weavers, shuttles seem to be in good condition after running nearly two years? That such is a fact no fixer will deny. It depends on the care they get. A set of new shuttles can be ruined in weaving out one warp, either by the shuttle striking the top of the box as it enters, or by striking the side. If the shuttle is worn flat, even if it be made smooth afterward, it will chafe the warp much more than if kept rounding.

Shellacing shuttles.—If the loom is breaking the shellac on the shuttles, it should be stopped at once. If the shellac is worn off you cannot afterward keep the wood smooth. Shellac should always be kept on hand, and whenever it is necessary to sand-paper the shuttles they should be covered with a coat before being used. Weavers should not be permitted to sand-paper their shuttles. It is only wearing them out. The right way to keep them smooth is to shellac them. In shellacing shuttles, remember that if they are coated with dust, covering them with shellac will not remove the dust, but, instead, will fix it on the shuttle. They must be cleaned and left perfectly smooth before applying the shellac.

Points of shuttles.—The points of the shuttles can be ground so as to feel smooth, and yet be in a condition to break out the thread of the warp.

They will do this when the fixer can hardly be made to believe it. He tries the shuttle by pushing the points through the threads, and if they slip off, all right; he thinks no damage can be done by breaking threads. The shuttle goes through much faster than the fixer can push it through, and if a thread comes in its way the shuttle will break it. To avoid all danger from this, grind the points so as to keep them well tapered. You will not expect to do this at one grinding, but each time they are ground the fixer should keep this in mind and maintain the shape of the tip.

Meeting in the shed.—One other effect of blunt shuttle-tips: they sometimes meet in the shed. On the 1880 loom this is more frequent than on others. If the tips are properly tapered they cause both of the shuttles to fly out of the shed if they meet. If the tips are blunt, they catch one another and have only mom ntum enough left to slide alongside of each other, thus causing a smash.

The points of new shuttles should be rounded a little before starting. The points are too sharp to use without some grinding. They should not be ground enough to make them blunt, but only to have the sharp point taken off. If this be not done, they make such a steep hole in the picker that the shuttles cannot slip out of it when dropping from one box to another and this may cause them to meet.

Injury done by shuttles.—In an ordinary sized room there are hundreds of threads broken every day by shuttles not being in proper condition. The threads thus broken are bad to sew in, as the end is usually carried as far as it will go, in the shed. Many of them are missed by both sewers and burlers and are only seen when the goods are finished. This, together with the trouble a poor set of shuttles gives the fixer, would seem to show that it is a thing of great importance.

Temples.—Time was when anything would do for a hook to hold the cloth out at the selvges. The contrivances used for this purpose when viewed in the light of present requirements, are varied and interesting. We can remember some first-class weave-rooms where a piece of wire, bent so as to form two hooks, was used. Some of them were so bent that the weight came entirely on one hook. A torn and ragged selvage is the result of using such hooks. Nor is this all; the cloth should be held out as near the full width as possible. The warp cannot weave well if the sides are being worn by the reed. It also makes the goods imperfect. Anything that serves to rough up the yarn should be fixed at once, for it makes a bad imperfection on some goods.

On some kinds of goods a temple cannot be successfully used, or at least the old style Dutcher temple cannot. The number of cases where a temple will not run is not so many as fixers suppose. It depends on how they are set and also on the judgment of the weaver.

The use of hooks.—We will suppose you are using hooks. They should be made with at least four hooks in each strap. To put the hooks in the leather as it can and should be done, is a very nice operation, and it requires not a little skill and patience to perform it. The hook should be made before you put it in the strap. One way of making a hook is to take a coarse file and rasp off the iron until it has somewhat of a taper. It is most likely to have a taper with three sides to it, like the point of a three-cornered file. If there are corners in the hook, where it is turned, they will surely tear the selvage so that they have to be avoided if you want to have the thing right.

The proper way is to finish tapering the wire with a smooth file and be careful to leave the wire round when you are done filing. The point should not be too slender. It can be made sharp enough to go in the selvage all right without weakening the point.

Before bending the hooks each one of them should be ground, not entirely for the purpose of sharpening them but to make the wire smooth, so that the inside of the hook will be smooth when bent. It is best to bend them with an old pair of round-nose plyers so as not to scratch the the wire.

Do not bend the hook so that the inside is N shaped. It is not fit for the purpose if so bent. It must be rounded nicely. The hooks should be put in the leather so that each one may be tight enough to be prevented from turning. The ends of the wire, if care be used, can be made to sink into the leather and not feel disagreeable to the weaver. Make each hook so that they come of even lengths. Of course you will taper them so as to give the proper slant to the strap; but the hooks should all take hold when hooked in the cloth. If properly made a hook will bear twice the weight without tearing the cloth that they otherwise would.

Straps.—Rollers to bear up the straps should always be used. The strap should never be put over the corner of the breast beam. The hooks pull directly from the reed and each time they are hooked up they make a thin place in the goods. In former times, when attention was called to this, the excuse would be offered that the temple hooks caused it, and it was supposed to be something that could not be helped. It has got to be helped now.

On goods that draw in very hard, two hooks on each side are necessary. It is a great deal better to use two hooks than it is to overload one in attempting to hold the cloth out, for the weight comes in one place and will make thin places in goods that are very heavy. In using two straps, provide rollers for each one of them. In place of having weights on each strap, it is better to use a spring on one of them.

The Dutcher temple.—On most of the work a Dutcher temple will run very well, and it is superior to hooks. There is no shading of the goods as with the hooks. One reason why temples do not meet with much favor in some mills is because the fixer is not obliged to use them, and consequently if one is put on and it does not run without much trouble, he takes it off and throws it aside where it may get rusty and unfit for use. It was the same way with the Tucker Stop-motion when they were put in different mills for trial. If they gave any trouble they were laid in the window, and the fixers never tried to master them until the time came when they had to. So it is with the temple. If the fixer is obliged to make them run he can generally find a way to overcome the difficulties that come in his way. A weaver should be taught how to handle them; a great deal of it rests with them. Some weavers get along all right with a temple, while others cannot start up again after having a pick-out.

Setting the temple.—The temple should be set so that the end of it stands within ¼ of an inch from the edge of the warp threads. If it is too near the end of the cloth, it is apt to tear the fabric. When you draw the lathe up, the catch should be set so as keep the end of the temple $\frac{1}{8}$ of an inch from the reed. Some set them so that the temple is $\frac{1}{2}$ or even $\frac{3}{4}$ of an inch from the reed. The effect of this is to let the cloth begin to slip out before the temple can get back to hold it. The head of the temple should be set only high enough to clear the race-board. It should also point downward towards the reed.

On a breast-beam that has been battered up by the points of shuttles and the like, you find it difficult to get a solid bearing for the stand. If packing is employed, it is sometimes put in so that the temple points upward, and the head of it may only just clear the race-board, yet the slant of the temple being just opposite to that of the race, when the lathe comes up far enough to beat in the pick, the head of the temple may be $\frac{1}{2}$ an inch from the race.

These things should not be, as the trouble with any temple is that it holds the cloth too high. Fasten the stand firmly on the frame. It will not do to have the temple rocking to and fro. I think it would be an improvement if the stand or bracket that supports the temple extended down so that a lag-screw could be put through the casting on the inside of the breast-beam.

Method of putting on temples.—I once had

102

to put temples on a lot of new looms, and not liking to drill holes through the plate on the breast-beam I got a thin piece of wood fastened to the regular iron bracket, taking care to give it the right pitch both downward and outward. The part of the pattern that was to come on the inside of the breast-beam was 12 inches long, and had a slot 10½ inches. From this I got my brackets cast, and I found them a great convenience. The temple was held in just the right place, and, of course, it could never slip. One lag-screw was all that was required, and the slot enabled me to move the temple in a few seconds.

Other temples.—There are other temples than that old standard, the Dutcher. So far as holding out the cloth is concerned the English temple is a good one. It has, in place of the old-style burr or roller, a series of brass rings spiral shaped. It also reaches farther into the cloth, and having more hold, retains the cloth in the temple. There are objections to it, among which I might note that the absence of any kind of spring on the temple makes it rather dangerous to use, for should the shuttle stop in the shed, and opposite the temple, it will very likely break the latter. It certainly will if the loom fails to protect. Still, this temple did more to supply · a long felt want than anything else in the field for a few years.

The Hardaker temple, on the same principle as the English temple, but improved in the bearings and stands gives us an article of real merit. As with the English temple the cloth is held out well, and there is very little danger of the shuttle catching it and breaking it.

I do not think there are any kinds of goods that this temple will not hold; and if the rolls are kept in order there will be no danger of the warp-threads in the cloth being cut by the pins.

There are other temples in use, but I have mentioned the principal ones, and perhaps the most popular. Temples are growing in favor, because they are becoming a necessity, and the overseer will do well to look into the merits of anything that comes along in the shape of a temple. Not only do they prevent shading of the goods, but anything that is in any way likely to wear up and go bad on the sides of the warp may be given the same chance as the middle.

The belt on the loom.—The belt on the loom is about as little understood as anything can well be, while nothing can be of more importance. Many fixers never go farther than to notice that a certain loom which has been running for some time all right, and which is noticeable for the smooth and easy pick it had with the belt moderately tight, suddenly gets into such a condition that it will not go at all. Everything seems the same, and is the same, yet the shuttle will not go across. The lug-strap is lowered, and yet it will not pick. Then the belt is tightened, and if enough is taken out the loom then picks all right; but the weaver can hardly start the loom up on account of the tight belt.

Instead of having to upset everything to make the loom run as it has run for some time, the belt should have been cared for in time. Whether you can see it or not, the reason for the loom making such a sudden change is that the condition of the belt is different, and once it begins to slip it gets roughed up and is made worse. Some looms run right along with the belt very slack, and if it is so, the weaver can handle the loom easily. It ships easily, and is an immense advantage to the weaver. Other looms cannot be made to run in that way. The prime cause is the belt. Of course, other things have been altered in an endeavor to make the loom run better, but the belt should first be attended to. It is not as easy a matter to get a belt in proper condition when once it is out, as it is to keep it right when you have it right.

Oiling the belt.—Belts need oiling often enough to keep them pliable. On the pliableness of a belt depends its adhesiveness. There is a mistaken idea that a rough surface will hold on the pulley better than a smooth surface.

Perhaps this is why a belt is sometimes scraped. If a composition of gum, dirt and flyings has accumulated on the belt the only thing to be done is to scrape it off; but scraping for any other purpose is fruitless. New belts are put on with the flesh or smooth side to the pulley, for the reason that the smooth side allows more of the belt surface to touch the pulley, air is entirely excluded from under the belt, and every part of it adheres to the pulley. If the belt becomes dry the surface will not as readily bend itself to the pulley. The surface becomes coated with a kind of dust, the result of the wear of the belt, therefore it loses its adhesiveness, and whenever a belt slips it wears the surface uneven. From this condition of things a long train of troubles can arise.

I do not think there is anything better for keeping belts in good condition than pure neat's foot oil. It answers the purpose.

Putting on the belt.—In putting a belt on, the butt of the scaff should strike the pulley first. If the point of the scaff or joint strikes the pulley first, the slipping of the belt will roll up the points. Do not slash off the end of your belt in a hap hazard way when you wish to cut it. A tri-square should always be used and the ends of the belts cut perfectly square. Punch an odd number of holes. On a loom-belt five holes are needed. One single hole is usually not enough for the lace to go through, unless you have a very large punch. If two holes are punched into one, they should be made so that the oblong hole is lengthwise of the belt. You will then save the effective strength of the belt. This punching is better done with an oval punch.

Lacing the belt.—In lacing the belt keep the back or pulley side straight. Draw your lace so that an equal portion of the strain is borne by each hole. Keep the edges straight.

In making a proper job of this belt-sewing, I do not think loom-fixers need instructions in sewing so much as they need to be impressed with the importance of the work.

It will not be denied that the work of mending belts in the weave-room is often hastily done and the resulting neglect is a thing of great moment where there are so many belts used.

Belt slipping.—It sometimes happens that the belt is inclined to run to one side of the pulley. The loom may start easy and be very hard to stop, or *vice versa*. Of course, if such is the case, the loom-shaft is not parallel with the line-shaft. The tendency of a belt is to the high part of a pulley if the shafts are parallel, but there is no high side to loom pulleys. They are straight across the surface, so that if the belt runs strongly to any particular side that side is nearest to the line-shaft. The belt will run to the ends of the shafts which are nearest together. This being known, if your belt does not run right, the weaver will have more comfort if the fixer will take a little trouble and fix it as it should be.

Running the loom.—Our loom should now be in good shape to start up and weave the warp out to the best advantage. We have touched upon everything that is necessary to success and if everything has been properly and thoroughly done there will be little trouble with the loom.

In running the loom the same principles must be followed as are laid down for a proper over-hauling. If the loom gets an ugly pick through the belt slipping, you cannot expect to keep it in good order by altering the picking-motion.

In saying that certain rules bring about certain results, it is taken for granted that other things are right, one part of the loom works conjointly with others, and each part must be working right to insure harmonious action of the whole. The habit that some fixers have of doing too much fixing in certain cases, is one of their greatest troubles. If a loom in its principal parts is in good condition, and has been running successfully and then begins to work badly, look out for some simple thing that has gone wrong. The whole loom cannot get upset so suddenly as that. Find out what it is before touching anything.

CHAPTER IX.

IN THE WEAVE-ROOM.

In the weave-room.—A large volume might be written touching upon all the details of loom-fixing, and the teachings of experience thus given would be a decided benefit, no doubt, more especially to beginners. There is, however, not room for the minutest details in this little treatise, the purpose of which, more than anything else, is to lead those who are engaged in the weave-room, whether overseers, fixers or learners, to think for themselves. It is a common habit for fixers to try to follow the customs of others, in the pursuit of their calling. I know that beginners have been the subject of ridicule on account of their seemingly awkward way of doing things. The reason they seemed awkward was because they were a little different from others in their methods. Now, I do not think it is necessary to be well up in the technical slang of a class of loom-fixers in order to meet with success. To know what you are trying to do is of paramount importance, and that is all you want of any one's ideas. Experience teaches you which way is the handiest to you in doing your work and that is as good as the way of anyone else. Do not be disturbed by those mysterious expressions indulged in by some. I remember a fixer who, in extolling his merits as a first-class hand, ran over a list of the looms he had worked on and capped it by saying, " I have also worked on the horizontal!" The manufacturer to whom he was talking, had never heard of the horizontal harness-motion loom and he thought he must be an extra good man.

The more a man learns on any subject, the more he realizes his own weakness and ignorance; and when a person gets just far enough to think that he is perfect because he has " worked " on this or that loom, he is in a pitiable condition.

Learners have been discouraged many a time by these persons, and I would expose the sham, and encourage them to brace up, and feel that their ideas are as good as any one else's. If they can only reach that point it would do them more good than anything else they can learn.

An intelligent man can fix on any loom, it matters not to him whether it be a Knowles, a Crompton, or what not, it is all the same to him.

Responsibility of the overseer.—The overseer cannot be indifferent to the methods employed by the fixers in doing their work. He is responsible for the condition of the room, and it is his duty to himself and his employers to see to it that due care is used in everything pertaining to the work. The fixer will not be as likely to appreciate the value of the supplies used in the weave-room as would the overseer. There is no department in the mill where extravagance counts up as rapidly as in the weave-room. It is due to the fixers themselves that the head of the room should know what is going on. If one fixer can run his section more economically than another, he is entitled to credit for it, and not to be classed the same as the fixer who may be careless and extravagant. Some fixers chafe under the restraint put upon them by certain overseers. In this they show poor judgment. They are neither friendly to themselves nor to their overseer. It is unreasonable to expect that the overseer shall assume all the responsibility and leave the work to be done in accordance with some one else's ideas. If he values his own interests, he will assert his right to have things done as he wants them, and the fixer who cannot acquiesce, does not understand his place.

Supplies for the weave-room.—In the matter of supplies for the weave-room the judgment of overseers seems to differ. Some get nearly everything made at home, and in my opinion lose thereby, in most cases, as the makers of supplies and repairs have better facilities for making a suitable article than home machine-shops unless there is a shop connected with the mill that is equipped for doing all kinds of work. But in mills that depend on outside shops for their repairs it is cheaper and far more satisfactory to get the supplies at the shops of the makers.

Take, for instance, the studs used about the loom. A picking-stud, perfect in every way, can be obtained for fifteen cents. The same for sweep and picking-stick studs. No local shops can make them for three times that sum. The same also may be said of castings. Everything fits and will give better satisfaction, and the room may as well be furnished with everything requisite for making repairs promptly, as to wait until you have a break-down and then pay as much for an article as would furnish three such. The costly and bungling plates frequently put on broken castings should be avoided by keeping a supply of new repairs on hand.

It cannot be said that this is extravagant, as it can be easily shown that it costs much more to patch things up. The place to save is in the usage the parts are subjected to while on the loom. If the parts are cared for, and the habit of battering and bruising castings stopped, there will be fewer repairs needed, and those should be of the best and ready for use whenever needed. The same may be said of sweep-straps and the like. A good article can be obtained and they will give better results in the long run than by using old belting, &c.

A supply of bolts should always be kept on hand. When one is needed it is needed badly, and the cost of furnishing the amount needed from time to time, if bought in small quantities, would furnish a good stock of all kinds.

These remarks are worthy of the careful attention of every overseer of weaving.

Examinations.—When the rod is taken off after starting the warps, the laps should always be looked over by the overseer or second hand. It is desirable that some one should do this who has an idea what the goods should look like. One person should see all the goods in the room. If all of them are seen by one person, he can more readily detect anything wrong or changed by the comparison he is enabled to make.

Besides looking over for wrong draws, the filling pattern should be just as carefully noticed. It is a very common occurrence for styles having a check, plaid, or figure of any kind to vary in size on different looms, of course nothing can be blamed for this but the picks, and for this reason some one should notice this at the start, and see to it that the picks are in and thus prevent this uneven weaving from the start. The filling pattern should be picked over, and also the harness-chain. This should, by all means, be done by some one besides the one who started it. His understanding of the draft may be wrong, and if he picked it over a dozen times he would probably read it just the same. A different person would be likely to see it in a different way, and thus prevent serious damages. None of these precautions should be omitted in any case.

Perching.—The cloth-perching in large mills has to be done by a percher, who gives his entire time to the work, sometimes requiring several assistants.

In moderate sized mills the overseer has to do his own perching. This works well enough in some cases, but it is often overdone. Manufacturers do not take into full consideration the importance of this branch of the weaver's work. A saving of ten times the expense of a percher might often be made in cases where the overseer has not sufficient time properly to perform his duties. The saving may be made, not alone in the mistakes discovered and made good before the cloth goes into the finishing-room, but also in preventing them from continuing in the pieces that follow.

If in the morning a piece be taken off containing a wrong draw, thread-out or other defect, the chances being that, whatever it was, it was still continuing, what an amount of sewing it will cause if the overseer should not get time to look that piece over till the latter part of the day? Even then he may hurry it through and over-look the mistake. All this is more than likely; it occurs very often. A piece should be looked over immediately after it is taken off. You may stop a wrong draw that has only just started, or perhaps one you have missed on the previous cut. Wrong draws and other like imperfections, continually going into the finishing-room will make trouble for somebody, and they should be stopped at any cost. It is better to lose a good deal with a loom standing than to lose a little by mistakes. The last heading should be carefully examined on every cut, and it is a good plan to teach the weavers to do this before hooking the cloth down. Many imperfections may be caught in this way.

Ticketing.—The tickets should be attached

to the last end of the piece and on the face. In order to insure regularity in this the weaver should always weave the string in the heading with the ticket attached. They should never be allowed to fasten the ticket to the cloth after it is taken off the loom, for they are liable to get it on the wrong end.

The pieces are finished towards the number end, and if one ticket should be put on one end, and on the next cut should be put on the opposite end, if the style was one that had a figure that had to be kept right side up, or a top and bottom to it, as it were, in order to keep them right, the nap would have to be put in opposite directions on these two pieces in cutting them.

Thorough inspection needed.—In inspecting the cloth over the perch, each piece should receive a thorough inspection. The piece should be looked over the face, looked through, and on most goods it pays to look the back over. On heavy weights this has to be done and it is sometimes useful on light weights, for an imperfection will show plainer on the back on some weaves than it does on the face. This is due to the twist of the yarn.

Measuring.—Everything should be marked before passing the piece over to the sewer-in. In measuring it must be remembered that it is not an easy thing to draw a piece over and come out exactly alike with a yard stick. It is easy to make a considerable percentage of difference in the lengths and this is an important matter, for it effects the weights. The finisher cannot make accurate calculations if there are variations in the weaver's weights and measurements. The measuring machine for perches is not a luxury entirely. By its use the exact lengths are obtained and the weights of the goods are more reliable. Variation in the weights should be followed up immediately after the piece is weighed. It will not do to wait until the next one comes off, as then you will be more likely to have two wrong weights instead of one.

The causes of uneven weights are pretty well understood. The way to prevent them will probably never be very satisfactorily known. The only way to obtain the best results is for the overseer to take every precaution that circumstances suggest and not wait until he is in trouble before beginning to look after the work. The lots should be numbered, and, whenever a new lot is started, its effect should be carefully noted so that any change that is necessary to be made, can be done before it is too late.

Sewing-in.—After the inspection and marking of the cloth it is passed to the sewer-in. This work if not in the control of the overseer of the weaving, should be, for none would derive as much benefit from a knowledge of the condition of the goods as he would. By having the direction of the sewing, he sees everything and can thus keep himself acquainted with the work of each weaver in particular, and the whole room in general. Under his direction the mending can be thoroughly done and according to the weave. The pieces should be looked over after the sewing has been done. It is the only way that the overseer can be sure that his work is complete and perfect.

Books for the weave-room.—The books for the weave-room are seldom alike in different mills. I have found good points in every kind that I have used. Their arrangement is generally made to suit the circumstances of the mill where the book is used. One way would never suit some mills, while in others it may be just what is wanted.

In some mills each piece is numbered consecutively. This is convenient for entering, for each piece is entered as it is measured; first come first served. But for reference it cannot be commended, neither is it convenient in making up the payroll. I think the most popular and convenient style of weave-book is like the form here given. There may be slight changes to suit different customs.

Date.	No.	Name.	Loom.	Picks	Shuttles.	Price.	Style.	Yds.	Weight.	Beam.	Lot.	Remarks.
Aug. 4. 6. 8.	49711	J. Benton.	19	64	3 P & P	10.3	911	32½ 31½ 33	26.8 26.4 26.5	380	720 720 732	
	12	"										
	13											
	14											
	15											
	16											
	17											
	18											
	19											
	49720											
Aug. 5.	21	Ella Mann.	15	42	2 2 & 2	7.5	820	31	24	378	530	Harness Skips.
	22											
	23											
	24											
	25											
	26											
	27											
	28											
	29											
	49730											
	31											
	32											
	33											
	34											

CHAPTER X.

CALCULATIONS.

CALCULATIONS FOR WOOLEN YARNS—"RUNS" AND "CUTS"
—FULL EXPLANATIONS—METHODS WITH WORSTED YARNS,
TABLE OF "RUNS," "CUTS," YARDS AND GRAINS—WEIGHT
IN A YARD OF WARP—FILLING CALCULATIONS—TO FIND
RUNS FROM OUNCES—POUNDS NEEDED FOR CUTS—PERCEN-
TAGES OF YARNS—SIZES OF PULLEYS—PERCENTAGES OF
WOOL, ETC.

Woolen-yarn calculations.—The base for
woolen yarn calculations is the "run." A run
of yarn is 1 pound spun to a length of 1600
yards. When spinners are paid by the run, they
receive so much for spinning 1600 yards of yarn
It is well to keep this in mind in making a study
of textile calculations.

Anyone can understand what you mean when
you say you want a 4-run yarn; but there are many
who would not understand you if you were to say
you wanted 600 runs of 4-run yarn. Therefore
to know what a run of yarn is becomes essen-
tial in making calculations. The use of the
terms "3-run," or "4-run" yarn might be

changed and the terms ⅛ or ¼ used. A 3-run thread is ⅓ the size of a 1-run thread, because in 1 pound cf 1-run yarn there are 1600 yards, in 1 pound of 3-run yarn there are 4800 yards; therefore the 3-run, being spun out to 3 times as many yards as the 1-run, is only ⅓ the size. 4-run yarn has 6400 yards per pound and is only ¼ the size of 1-run yarn. So if we want to add two or more threads together, as in making double and twist yarn, we can treat them as we would fractions, nearly.

To add together the fractions ½ + ½ we know it would make one whole. Thus, $\frac{1}{2} + \frac{1}{2} = \frac{1}{1} = 1$. Now, of course you can readily see that adding 2 2-run threads together makes a 1-run thread, so that in this case, it reads right. But we will take another. Add ⅓ and ¼ together —in other words, add a 3-run and a 4-run together

$$\frac{1}{3} + \frac{1}{4} = \frac{7}{12}$$

Now we know that these two threads cannot make one as heavy as $\frac{7}{12}$ of a run, less than $\frac{3}{4}$. In adding two fractions having 1 for their numerator, the rule is:

Multiply the denominators together for a new denominator, add them together for a numerator.

This we have done. The result is the fraction $\frac{7}{12}$, which means that it is not as *heavy* as the 1-run thread. So the fraction expresses correctly the size of our thread, taking 1-run for a base. But we want to have it expressed in terms that are more pertinent to the subject. This we can do in all cases by first inverting the terms, and then proceeding as in ordinary fractions. Thus: $\frac{12}{7} = 1\frac{5}{7}$ runs, which is correct.

Another example: What weight of thread have we by putting together 3, 7 and 9 runs

$$\frac{1}{3} \times \frac{1}{7} \times \frac{1}{9}$$

$$
\begin{array}{l|l}
 & 63 \\
\hline
3 & 21 \times 1 = 21 \\
7 & 9 \times 1 = 9 \\
9 & 7 \times 1 = 7 \\
\hline
 & 37 \\
\end{array}
$$

$$\frac{37}{63} = \frac{63}{37} = 1\frac{26}{63}$$

or 1½ run heavy.

There are other ways given for doing this, but I know of no better way, and it is one that will put before you in plain terms the size of your doubled thread.

Cuts and runs.—The yarn table given herewith is for runs and cuts. There are many mills where yarn is numbered by cuts entirely. To any one accustomed to numbering by runs it is confusing.

A cut of yarn contains 300 yards. In 3-run

yarn we have 4800 yards in a pound. It will take 16 cuts to equal a 3-run yarn; 24 cuts multiplied by 300 equals 7200 yards. The latter sum divided by 1600 gives us 4½ runs. Hence to convert cuts into runs, divide the yards per pound in the cut numbers by 1600. To convert cuts into runs, divide the yards in the run numbers by 300, and you have the cuts.

Worsted yarns.—In worsted yarn there are 560 yards per pound, this is called a hank. Worsted is used in woolen goods so much now that it is quite necessary for those employed in the woolen business to understand the relative value of the worsted thread in weight as compared with a woolen thread.

You proceed in the same way as in cuts. A 5-run thread contains $5 \times 1600 = 8000$, $8000 \div 560 = 14.28+$ which is a worsted number. A No. 20 worsted thread is $560 \times 20 = 11,200$ yards, $11,200 \div 1600 = 7$ runs. Worsted yarn is used double on all except the heavy numbers, as 2-50s, 2-40s, &c. The meaning of this is that two No. 40 worsted threads are put together, so that the two are twice the size of a single thread, which will make them the weight of a No. 20, 2-50s would be the weight of a No. 25, &c.

All yarns are calculated on the basis of 7000 Troy grains in 1 pound avoirdupois. The object is to have the avoirdupois pound, and to use the

Troy grain as convenient divisions of it. That is all there is to it. There are those who have different views on the subject, but the best authorities agree that the above is the right way.

TABLE OF RUNS, CUTS, YARDS AND GRAINS.

Runs.	Cuts.		25 yds. Grains.	50 yds. Grains.	Runs.	Cuts.		25 yds. Grains.	50 yds. Grains.
½	2⅔		218.8	437 5	6¾	36		16.2	32.4
1	5⅓		109 4	218.8	7	37½		15.6	31.3
1¼	6⅔		87 5	175.	7¼	38⅔		15 1	30.2
1½	8		72.9	145.8	7½	40		14 6	29.2
1¾	9⅓		62 5	125.	7¾	41⅓		14.1	28.2
2	10⅔		54 7	109.4	8	42⅔		13 7	27 4
2¼	12		48.6	97.2	8¼	44		13.3	26.5
2½	13⅓		43.8	87.5	8½	45⅓		12.9	25.7
2¾	14⅔		39 8	79 5	8¾	46⅔		12 5	25.
3	16		36.5	72.9	9	48		12.2	24.3
3¼	17½		33.7	67.3	9¼	49⅓		11.8	23.6
3½	18⅔		31.3	62.5	9½	50⅔		11.5	23.
3¾	20		29.2	58.3	9¾	52		11 2	22.
4	21⅓		27 3	54.7	10	53⅓		10.9	21.9
4¼	22⅔		25.7	51 5	10¼	54⅔		10.7	21.3
4½	24		24.3	48.6	10½	56		10.4	20.8
4¾	25⅓		23.	46.1	10¾	57⅓		10.2	20.4
5	26⅔		21.9	43.8	11	58⅔		9.9	19.9
5¼	28		20.8	41 7	11¼	60		9 7	19.4
5½	29⅓		19.9	39.8	11½			9.5	19.
5¾	30⅔		19.	38.1	11¾			9.3	18.6
6	32		18.2	36 5	12			9 1	18.2
6¼	33⅓		17.5	35.					
6½	34⅔		16 8	33.7					

Weight in a yard of warp.—To show how to calculate the weight of yarn in a yard of warp, we will take a warp containing 1600 ends of 4-run yarn. 1 yard of warp gives us 1600 yards of thread. If that 1600 yards were 1-run yarn, we should just have 1 pound of warp. Being 4-run, which is ¼ the size of the 1-run, we have ¼ of the weight, ¼ + 16 = 4 oz. So to obtain the weight of yarn in the warp, divide the threads by the runs.

Example: 400)1600
           4.00 oz.

Another: 5.25)4200(8 oz.
           4200

Write the runs decimally for convenience in case you have fractions of runs, as 5.25, 5.75, &c.

Weight of yarn.—To obtain the proper weight of yarn, the weight of the goods wanted being known, the process is just opposite to the other. To illustrate: We want to put in 1600 threads and want the warp to weigh 4 oz., what weight shall we spin the yarn?

           oz. 400)1600
               400 runs.

or,   800)4200(5.25 runs.
           4000
           200.0
           1600
           400.0

Filling calculation.—For the filling we multiply the picks per inch and the width in the reed together, to get the yards in the filling. This may not appear quite clear to the learner. It looks strange that 1 inch of filling multiplied by the width of the cloth, equals all the threads in the warp. Well, suppose we take a strip of the cloth 1 inch wide, and we go lengthwise for 1 yard, 36 inches. There are 46 picks per inch. so $36 + 46 = 1656$ inches of yarn. Now, we have only got 1 inch of the width and 1 yard of the length. We multiply 1656 by the total width, which we will call 36 inches. We then have 59,616 total inches of filling in 1 yard of cloth.

We then divide by 36 to get these inches into yards and we have 1656 again, hence, the simple rule.

Multiply the width in the reed by the picks in 1 inch for the yards of filling in 1 yard of cloth.

Example: runs, 4)1656
4.14 oz.

To find the runs from ounces.—To find the runs, the ounces being known, divide the threads by ounces instead of the runs. The weights thus obtained are the weights off the loom, the yarn being exact, and no account being taken of the listing, shrinkage in weaving, &c. But these will count up, of course, and it will be found that the goods will be heavier than the weights produced by the calculation. This, together with the shrinkage in length in the finishing, will compensate for the loss in weight by scouring, gigging and shearing, &c. So that the weights finished will correspond with the weights given by the calculation as near as can be. There are those who may be more elaborate in their method, but the results are no nearer correct, of this I am certain.

Pounds needed for cuts.—To calculate the amount of yarn required for a considerable quantity of yarn, we proceed a little differently. The rules we have just given relate to the weight per yard. We wish to find the pounds of yarn needed for 1 or more cuts.

$$\begin{array}{r} 2970 \text{ threads of 4-run warp.} \\ 40 \text{ yards per cut.} \\ \hline \end{array}$$

1 run of yarn, 1600)118800(74.25 runs.

By the above process we multiply the threads by the number of yards it will require to weave a cut of cloth. The cut may be 35 yards or a little more. We allow it 40 yards of yarn for take-up in weaving. We have 118,800 yards of yarn, 1 run of yarn (1600) is contained in that 74.25 times, so we have 74.25 runs. To get this into pounds, divide by the size of the yarn, thus:

run 4)74.25
18.56¼ pounds of warp

for 1 cut. To get the filling, we proceed as in the former examples.

$$
\begin{array}{r}
65 \text{ inches.} \\
66 \text{ picks.} \\
\hline
390 \\
390 \\
\hline
4290 \\
35 \text{ the yards of cloth.} \\
\hline
21450 \\
12870 \\
\hline
\end{array}
$$

1600)150150(93.84 runs.
$$
\begin{array}{r}
1440 \\
\hline
615 \\
480 \\
\hline
1350 \\
1280 \\
\hline
700
\end{array}
$$

We have multiplied by 35 the actual length of the cut when woven. We make no allowance for take-up, because the take-up in the warp does not affect the amount of filling put in.

Size of yarn, 4.25)93.84(22.08 lbs. of filling.
$$
\begin{array}{r}
850 \\
\hline
884 \\
850 \\
\hline
3400
\end{array}
$$

The weaver can make an estimate of the filling required to take out certain warps that are in the loom in this convenient way. Suppose you have

21 cuts in the looms, there are 56 picks of 4.25 run, 74 inches wide:

$$
\begin{array}{r}
74 \\
56 \\
\hline
444 \\
37.0 \\
\hline
4144 \\
35 \quad \text{yards.} \\
\hline
20720 \\
12432 \\
\hline
\end{array}
$$

1600)145040(90.65 runs.
1440
————
1040

90.65 run, multiplied by 21 (the cuts), gives 1903.63, total runs wanted.

Percentages of yarns.—In making calculations on the percentages of yarns required where different kinds are used in one warp the following examples and illustrations will be of benefit to some, I think.

You have a warp dressed as follows:
2 threads of 4-run yarn.
1   "   2   "

What percentage of each one is required?

We will take the lightest thread, and take, say, one pound for a basis. You could take 10 or 100 pounds just as well, but this will do. By taking the lightest thread we are sure that the others will weigh more than one pound and the point will be easier to see. We say a certain amount of 4-run yarn weighs 1 pound. We have two 4-run threads, so we repeat that and set another pound down, under the first. We then have the 2-run thread and we know that it weighs twice as heavy as the 4-run thread without calculating. But, if it is not so plain in other sums that may happen we obtain the right result by dividing the 4-run by the 2-run thread. This gives us 2 pounds. Add them all together and we have 4 pounds, the total weight of the three threads.

Now, we want to find the percentage of each one of them and we can then make our batches to suit the quantity of yarn of each kind that we need. You find the percentage just the same as you would find what percentage you had taken from $1.00 if you had taken 10 cents away from it, which would be done in the following simple way:

$$\$1 \mid 10c.$$
$$10 \text{ per cent.}$$

We add two cyphers to the sum subtracted, and divide by the original sum. The quotient is the per cent. taken from $1.00.

To find the percentage of each kind of yarn we proceed in the same way.

Write the pounds decimally as in some instances you will find it necessary.

1 4-run thread=1.00 pound
1   "   "   =1.00   "
1 2-run   "   =2.00   "
                ————
                4.00

Now, we get the percentage of each one and from that we can always make a calculation as to the amount of yarn wanted.

The operation :

1 4-run=1.00
1   " =1.00
1 2-run=2.00
        ————
        4.00 lbs.

4 | 100
    25 per cent. of 1 4-run thread.

4 | 200
    50 per cent. of 1 2-run thread.

1 4-run thread=25 per cent.
1   "   "   =25   "
1 2-run   "   =50   "
                ————
                100

Proceed in the same way no matter how many threads there are, making a separate item of each thread, the percentage of all threads that are alike can be added together afterward. As in this illustration we have 25 per cent. for each 4-run thread, they are each of one kind of stock so we put them together and have 50 per cent.

To calculate the percentage on wool.—

We have a mix composed of

> 70 per cent. black,
> 20 "   "   blue,
> 10 "   "   white.

You have on hand, say 165 pounds of black. You want to know how many pounds of each of the other colors to use to make up the proper proportion, so that you can use all of the black you have on hand, how would you go about it?

Some would say 20 per cent. of 165 is 33. 10 per cent of 165 is 16.50.

Let us see if this would be right. We will add them together.

> 165   black,
> 33 · blue,
> 16.50 white.
> ———
> 214.50 total batch.

70 per cent. of this sum should be 165, for, whatever amount of the other colors are used, the quantity of black on hand must be 70 per cent. of the whole batch. 70 per cent. of 214.50 is 150.14. We have 150 pounds as representing the 70 per cent. of black, while we have put in 165 pounds. You will see at once that it is wrong; the reason why it is wrong is that we have taken 20 per cent. and 10 per cent., respectively, of what at the start was only 70 per cent. of

what the whole should be, thus lowering its percentage and increasing the rest of them.

. Now, let us try another way. If 165 is 70 per cent. of the batch wanted, what is 1 per cent. of it? If we can get that we can multiply the 1 per cent by 10 or 20 per cent or any other amount, and it gives us the right result each time. To get 1 per cent. we divide 165 by 70, thus:

$$
\begin{array}{r}
70)165(2.357 \\
140 \\
\hline
250 \\
210 \\
\hline
400 \\
350 \\
\hline
500 \\
490 \\
\hline
10
\end{array}
$$

Having obtained 1 per cent. we multiply it by each of the proportions wanted.

Black—70 times	2.357=	164.990	℔s.
Blue —20 "	"	47.140	"
White—10 "	"	23.570	"
		235.700	

It will be seen by the above that we have for black 164 $\frac{99}{100}$ pounds, which is as near right as it can be brought. To prove the work, add the 10 and 20 per cent. together and it should leave 164.99, when those two are subtracted from the the total batch.

By careful study of the principles involved in these examples of textile calculation, anyone endeavoring to learn can find why these problems are worked out the way they are. I have endeavored to avoid mysterious signs and terms, remembering the remark made by a young man who aspired to learn but " got discouraged by those crosses, dots and signs." To those who know something more of mathematical calculations than others, these examples will be none the less plain.

Sizes of pulleys.—We have a loom with a 12-inch driving pulley making 2 12/17 revolutions to each pick, the line-shaft makes 146 revolutions per minute, what size of pulley do we want to run the loom 80 picks per minute?

2 12/17 × 80 = 216.47, the revolutions per minute of the loom pulley. 216.47 × 12 = 2597.64 ÷ 146 = 17.10 size of pulley.

Multiply 2 12/17 the revolutions per pick, by the picks you want the loom to run, this gives speed of the loom pulley.

Multiply this by size of the loom pulley 12-inch, and the product divided by the speed of the line-shaft, 146, gives the size of the pulley we want, 17.10.

THE END.

www.ingramcontent.com/pod-product-compliance
Lightning Source LLC
Chambersburg PA
CBHW021818190326
41518CB00007B/652